SOLUTIONS MANUAL

MICROWAVE TRANSISTOR AMPLIFIERS

Analysis and Design

SECOND EDITION

Guillermo Gonzalez

PRENTICE HALL
Upper Saddle River, NJ 07458

Acquisitions Editor: *Eric Svendsen*
Project Editor: *Lorena Cerisano*
Production Manager: *Joan Eurell*
Production Coordinator: *Julia Meehan*

A Pearson Education Company
Upper Saddle River, NJ 07458

Printed in the United States of America

10 9 8 7 6 5 4 3 2 1

ISBN 0-13-271255-5

Prentice-Hall International (UK) Limited, London
Prentice-Hall of Australia Pty. Limited, Sydney
Prentice-Hall Canada Inc., Toronto
Prentice-Hall Hispanoamericana, S.A., Mexico
Prentice-Hall of India Private Limited, New Delhi
Prentice-Hall of Japan, Inc., Tokyo
Pearson Education Asia Pte. Ltd., Singapore
Editoria Prentice-Hall do Brasil, Ltda., Rio De Janeiro

CONTENTS

CHAPTER 1

1.1) IN (1.3.41) LET $\Gamma_0 = |\Gamma_0| e^{j\psi_0}$:

$$|V(d)| = |A_1| \, |1 + |\Gamma_0| e^{j(\psi_0 - 2\beta d)}|$$

THE MAXIMUM VALUE OF $|V(d)|$ OCCURS WHEN $e^{j(\psi_0 - 2\beta d)} = 1$, AND THE MINIMUM VALUE OCCURS WHEN $e^{j(\psi_0 - 2\beta d)} = -1$. HENCE,

$$|V(d)|_{max} = |A_1| \, |1 + |\Gamma_0|| = |A_1|(1 + |\Gamma_0|) \quad \text{AND} \quad |V(d)|_{min} = |A_1| \, |1 - |\Gamma_0|| = |A_1|(1 - |\Gamma_0|)$$

1.2) $v(d,t) = \text{Re}[V(d) e^{j\omega t}]$

$$V(d) = 3.95 \, e^{j(\beta d - 63.44^\circ)} + 1.77 \, e^{-j\beta d}$$

$$\therefore \; v(d,t) = 3.95 \cos(\omega t + \beta d - 63.44^\circ) + 1.77 \cos(\omega t - \beta d)$$

AND $i(d,t) = \dfrac{3.95}{50} \cos(\omega t + \beta d - 63.44^\circ) - \dfrac{1.77}{50} \cos(\omega t - \beta d)$

1.3) (a) $\Gamma_0 = \dfrac{Z_L - Z_0}{Z_L + Z_0} = \dfrac{(100 + j100) - 50}{(100 + j100) + 50} = 0.62 \angle 29.7^\circ$

$$Z_{IN}\left(\tfrac{\lambda}{8}\right) = 50 \, \frac{(100 + j100)\cos 45^\circ + j50 \sin 45^\circ}{50 \cos 45^\circ + j(100 + j100)\sin 45^\circ} = 40 - j70 \ \Omega$$

$$\text{VSWR} = \frac{1 + |\Gamma_0|}{1 - |\Gamma_0|} = \frac{1 + 0.62}{1 - 0.62} = 4.26$$

(b)

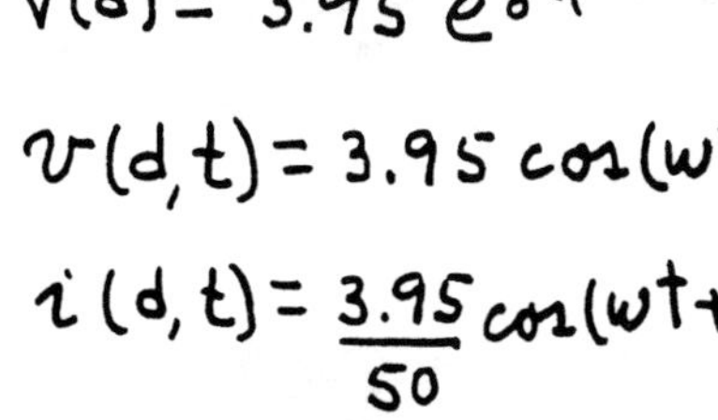

$$V\left(\tfrac{\lambda}{8}\right) = \frac{10\angle 0^\circ (40 - j70)}{40 - j70 + 100} = 5.15 \angle -33.7^\circ$$

$$I\left(\tfrac{\lambda}{8}\right) = V\left(\tfrac{\lambda}{8}\right) / Z_{IN}\left(\tfrac{\lambda}{8}\right) = 0.064 \angle 26.6^\circ$$

$$P\left(\tfrac{\lambda}{8}\right) = \tfrac{1}{2} \text{Re}\left[V\left(\tfrac{\lambda}{8}\right) I^*\left(\tfrac{\lambda}{8}\right)\right] = 82 \text{ mW}$$

$$V\left(\tfrac{\lambda}{8}\right) = 5.15 \angle -33.7 = A_1 e^{j\frac{\pi}{4}} \left[1 + 0.62 \angle 29.7^\circ e^{-j\frac{\pi}{2}}\right]$$

$$\therefore \; A_1 = 3.643 \angle -56.31^\circ$$

$$V(d)=3.643\angle{-56.31^\circ}\, e^{j\beta d}\left[1+0.62\angle{29.7^\circ}\, e^{-j2\beta d}\right]$$

$$V(0)=3.643\angle{-56.31^\circ}\left[1+0.62\angle{29.7^\circ}\right]=5.72\angle{-45.02^\circ}$$

$$I(0)=\frac{5.72\angle{-45.02^\circ}}{100+j100}=0.04\angle{-90^\circ}$$

$$P(0)=\frac{1}{2}\operatorname{Re}\left[V(0)I^*(0)\right]=82\ \text{mW}$$

AS EXPECTED: $P(\frac{\lambda}{8})=P(0)$

1.4) $b_1=S_{11}a_1+S_{12}a_2$

$b_2=S_{21}a_1+S_{22}a_2$

LETTING $Z_0=Z_{01}=Z_{02}$, THEN $b_1=\frac{V_1^-}{\sqrt{Z_0}}$, $b_2=\frac{V_2^-}{\sqrt{Z_0}}$, $a_1=\frac{V_1^+}{\sqrt{Z_0}}$, AND $a_2=\frac{V_2^+}{\sqrt{Z_0}}$.

$$\therefore\ \frac{V_1^-}{\sqrt{Z_0}}=S_{11}\frac{V_1^+}{\sqrt{Z_0}}+S_{12}\frac{V_2^+}{\sqrt{Z_0}}\Rightarrow V_1^-=S_{11}V_1^++S_{12}V_2^+$$

$$\frac{V_2^-}{\sqrt{Z_0}}=S_{21}\frac{V_1^+}{\sqrt{Z_0}}+S_{22}\frac{V_2^+}{\sqrt{Z_0}}\Rightarrow V_2^-=S_{21}V_1^++S_{22}V_2^+$$

WITH $b_1=\sqrt{Z_0}\,I_1^-$, $b_2=\sqrt{Z_0}\,I_2^-$, $a_1=\sqrt{Z_0}\,I_1^+$, $a_2=\sqrt{Z_0}\,I_2^+$

WE OBTAIN: $I_1^-=S_{11}I_1^++S_{12}I_2^+$

$I_2^-=S_{21}I_1^++S_{22}I_2^+$

1.5)
$$\begin{cases}a_1=T_{11}b_2+T_{12}a_2 & (1)\\ b_1=T_{21}b_2+T_{22}a_2 & (2)\end{cases}\qquad \begin{cases}b_1=S_{11}a_1+S_{12}a_2 & (3)\\ b_2=S_{21}a_1+S_{22}a_2 & (4)\end{cases}$$

FROM (3) AND (4):

$$a_1 = \frac{1}{S_{11}} b_1 - \frac{S_{12}}{S_{11}} a_2 \qquad (5)$$

$$a_1 = \frac{1}{S_{21}} b_2 - \frac{S_{22}}{S_{21}} a_2 \qquad (6)$$

EQUATING (5) AND (6):

$$\frac{1}{S_{11}} b_1 - \frac{S_{12}}{S_{11}} a_2 = \frac{1}{S_{21}} b_2 - \frac{S_{22}}{S_{21}} a_2$$

$$\therefore \quad b_1 = \frac{S_{11}}{S_{21}} b_2 + \left[S_{12} - \frac{S_{22}S_{11}}{S_{21}}\right] a_2 \qquad (7)$$

COMPARING (2) AND (7):

$$T_{21} = \frac{S_{11}}{S_{21}} \quad \text{AND} \quad T_{22} = S_{12} - \frac{S_{11}S_{22}}{S_{21}}$$

EQUATING (3) AND (7) GIVES

$$S_{11}a_1 + S_{12}a_2 = \frac{S_{11}}{S_{21}} b_2 + \left[S_{12} - \frac{S_{11}S_{22}}{S_{21}}\right] a_2$$

$$\therefore a_1 = \frac{1}{S_{21}} b_2 - \frac{S_{22}}{S_{21}} a_2 \qquad (8)$$

COMPARING (1) AND (8):

$$T_{11} = \frac{1}{S_{21}} \quad \text{AND} \quad T_{12} = -\frac{S_{22}}{S_{21}}$$

SIMILARLY, STARTING WITH (1) AND (2), IT FOLLOWS THAT:

$$\begin{bmatrix} S_{11} & S_{12} \\ S_{21} & S_{22} \end{bmatrix} = \begin{bmatrix} \dfrac{T_{21}}{T_{11}} & T_{22} - \dfrac{T_{12}T_{21}}{T_{11}} \\ \dfrac{1}{T_{11}} & -\dfrac{T_{12}}{T_{11}} \end{bmatrix}$$

1.6) THE OPEN-CIRCUIT VOLTAGE $E_{1,TH}$ IS OBTAINED AS FOLLOWS:

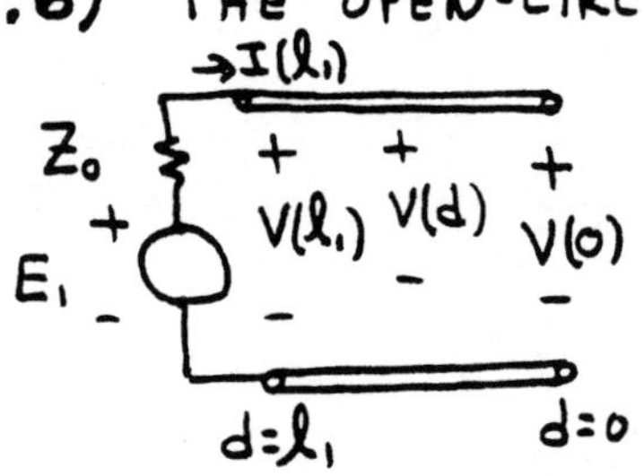

$$\Gamma_0 = 1 \qquad V(d) = A(e^{j\beta d} + e^{-j\beta d}) = 2A\cos\beta d$$

$$I(d) = j2\frac{A}{Z_0}\sin\beta d$$

$$V(\ell_1) = 2A\cos\beta\ell_1 = E_1 - I(\ell_1)Z_0$$

$$= E_1 - j2A\sin\beta d$$

$$\therefore A = \frac{E_1}{2(\cos\beta\ell_1 + j\sin\beta\ell_1)} = \frac{E_1}{2} e^{-j\beta\ell_1}$$

HENCE: $E_{1,TH} = V(0) = 2A = E_1 e^{-j\beta\ell_1}$

1.7) (a)

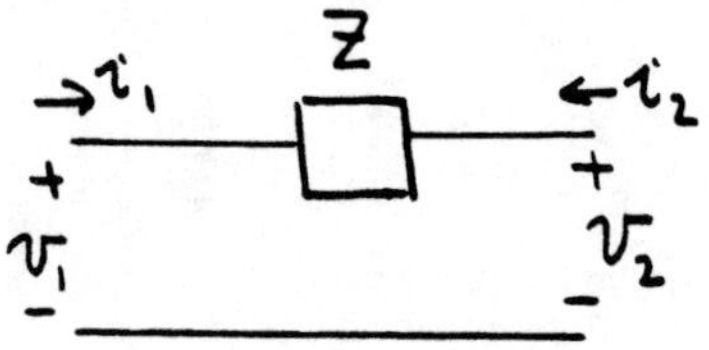

$$v_1 = A v_2 - B i_2$$
$$i_1 = C v_2 - D i_2$$

$$A = \left.\frac{v_1}{v_2}\right|_{i_2=0} \qquad \text{WITH } i_2 = 0: \quad v_1 = v_2$$
$$\therefore A = 1$$

$$B = -\left.\frac{v_1}{i_2}\right|_{v_2=0} \qquad \text{WITH } v_2 = 0: \quad v_1 = i_1 Z = -i_2 Z$$
$$\therefore B = -\frac{(-i_2 Z)}{i_2} = Z$$

$$C = \left.\frac{i_1}{v_2}\right|_{i_2=0} \qquad \text{WITH } i_2 = 0: \quad i_1 = -i_2 = 0$$
$$\therefore C = 0$$

$$D = \left.\frac{i_1}{-i_2}\right|_{v_2=0} \qquad \text{WITH } v_2 = 0: \quad i_1 = -i_2$$
$$\therefore D = 1$$

(b) FROM FIG. 1.8.1:

$$S_{11} = \frac{A' + B' - C' - D'}{A' + B' + C' + D'} = \frac{1 + \frac{Z}{Z_0} - 0 - 1}{1 + \frac{Z}{Z_0} + 0 + 1} = \frac{Z}{Z + 2Z_0}$$

$$S_{12} = \frac{2(A'D' - B'C')}{A' + B' + C' + D'} = \frac{2(1 - 0)}{1 + \frac{Z}{Z_0} + 0 + 1} = \frac{2Z_0}{Z + 2Z_0}$$

$$S_{21} = \frac{2}{A' + B' + C' + D'} = \frac{2}{1 + \frac{Z}{Z_0} + 0 + 1} = \frac{2Z_0}{Z + 2Z_0}$$

$$S_{22} = \frac{-A' + B' - C' + D'}{A' + B' + C' + D'} = \frac{-1 + \frac{Z}{Z_0} - 0 + 1}{1 + \frac{Z}{Z_0} + 0 + 1} = \frac{Z}{Z + 2Z_0}$$

FOR THE SHUNT ADMITTANCE Y THE ABCD MATRIX IS:

$$A = 1, \quad B = 0, \quad C = Y, \quad \text{AND } D = 1$$

AND

$$S_{11} = \frac{A' + B' - C' - D'}{A' + B' + C' + D'} = \frac{1 + 0 - YZ_0 - 1}{1 + 0 + YZ_0 + 1} = \frac{-YZ_0}{2 + YZ_0}$$

$$S_{12} = \frac{2(A'D' - B'C')}{A' + B' + C' + D'} = \frac{2(1 - 0)}{1 + 0 + YZ_0 + 1} = \frac{2}{2 + YZ_0}$$

ALSO: $S_{22} = S_{11}$ AND $S_{21} = S_{12}$

1.8) IN A Z_0 SYSTEM:

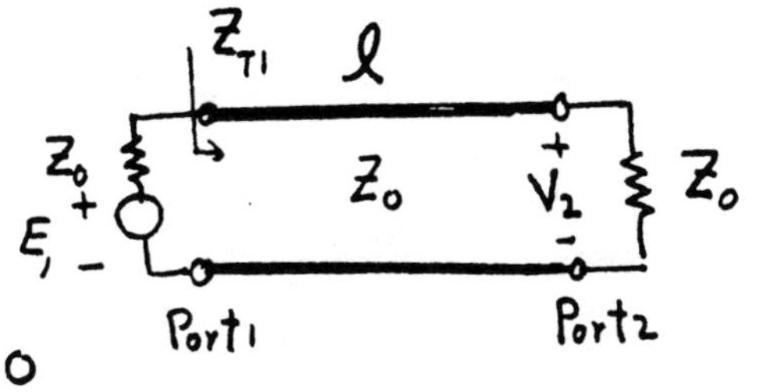

$$Z_{T1} = Z_0$$

$$S_{11} = \frac{Z_{T1} - Z_0}{Z_{T1} + Z_0} = \frac{Z_0 - Z_0}{Z_0 + Z_0} = 0$$

ALSO (FROM SYMMETRY): $S_{22} = 0$

$$S_{21} = 2\frac{\sqrt{Z_0}}{\sqrt{Z_0}}\frac{V_2}{E_1} = 2\frac{V_2}{E_1}$$

SINCE $V(x) = \frac{E_1}{2} e^{-j\beta x}$

$$V_2 = V(\ell) = \frac{E_1}{2} e^{-j\beta\ell}$$

$$\therefore \; S_{21} = \frac{2\left(\frac{E_1}{2} e^{-j\beta\ell}\right)}{E_1} = e^{-j\beta\ell}$$

ALSO (FROM SYMMETRY): $S_{12} = e^{-j\beta\ell}$

AN ALTERNATE WAY OF DERIVING S_{21} IS:

$$S_{21} = \left.\frac{b_2}{a_1}\right|_{a_2=0}$$

SINCE b_2 IS THE WAVE AT PORT 2 AND a_1 IS THE WAVE AT PORT 1, WE HAVE:

$$b_2 = a_1 e^{-j\beta\ell} \quad (a_2 = 0)$$

$$\therefore \; S_{21} = \frac{b_2}{a_1} = e^{-j\beta\ell}$$

$$T = \begin{bmatrix} \frac{1}{S_{21}} & -\frac{S_{22}}{S_{21}} \\ \frac{S_{11}}{S_{21}} & S_{12} - \frac{S_{11}S_{22}}{S_{21}} \end{bmatrix} = \begin{bmatrix} e^{j\beta\ell} & 0 \\ 0 & e^{-j\beta\ell} \end{bmatrix}$$

1.9)

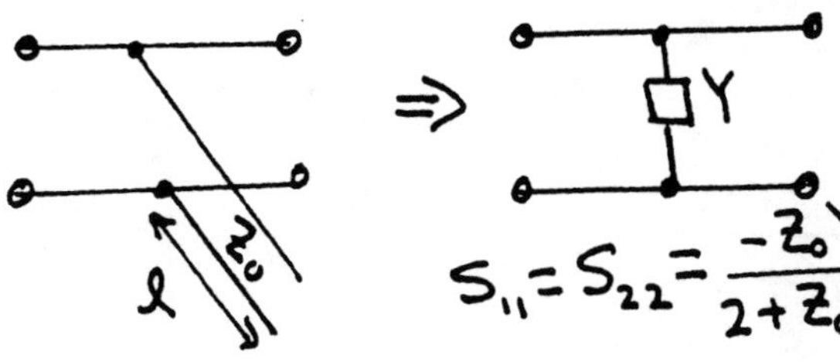

$$Y = \frac{1}{Z_{oc}} = \frac{1}{-jZ_0 \cot\beta\ell} = jY_0 \tan\beta\ell$$

$$S_{11} = S_{22} = \frac{-Z_0 Y}{2 + Z_0 Y} = \frac{-1}{1 - j2\cot\beta\ell}$$

$$S_{12} = S_{21} = \frac{2}{2 + Z_0 Y} = \frac{2}{2 + j\tan\beta\ell}$$

USE (1.4.11) TO EVALUATE THE T PARAMETERS.

1.10)

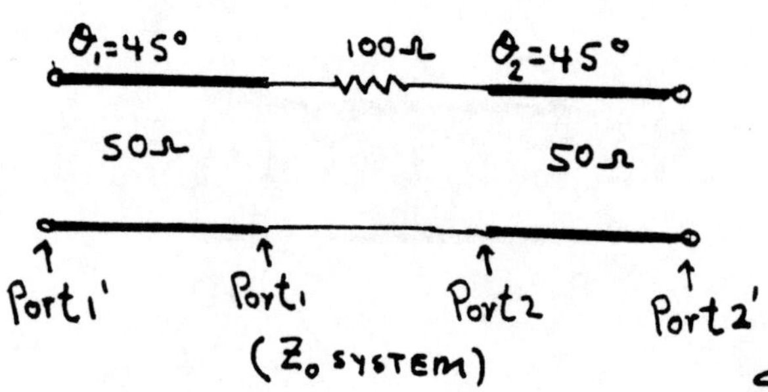

THE S PARAMETERS AT PORTS 1 AND 2 ARE:

$$S_{11}=\frac{Z}{Z+2Z_0}=\frac{100}{100+2(50)}=0.5$$

$$S_{22}=S_{11}=0.5$$

$$S_{12}=S_{21}=\frac{2Z_0}{Z+2Z_0}=0.5$$

THE S PARAMETERS AT PORTS 1' AND 2' ARE OBTAINED USING (1.5.4):

$$S'_{11}=S'_{22}=S_{11}e^{-j2\theta_1}=0.5\angle{-90°}=-j0.5$$

$$S'_{21}=S'_{12}=S_{21}e^{-j(\theta_1+\theta_2)}=0.5\angle{-90°}=-j0.5$$

1.11)

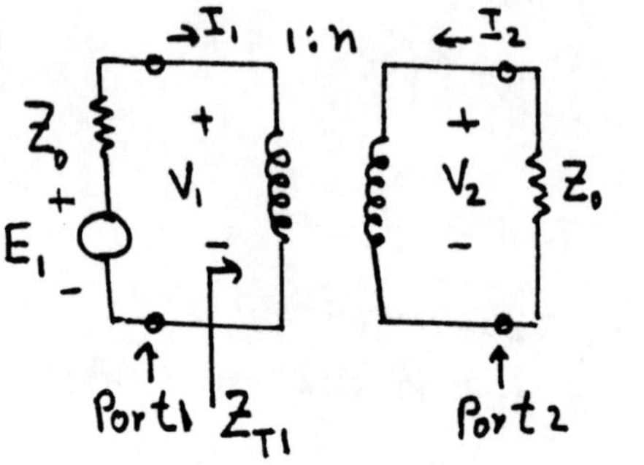

IN THE Z_0 SYSTEM SHOWN, THE PARAMETERS S_{11} AND S_{21} ARE CALCULATED AS FOLLOWS:

$$I_2=\frac{I_1}{n},\quad V_2=V_1 n$$

$$Z_{T1}=\frac{V_1}{I_1}=\frac{V_2}{I_2 n^2}=\frac{Z_0}{n^2}$$

$$S_{11}=\left.\frac{b_1}{a_1}\right|_{a_2=0}=\frac{Z_{T1}-Z_0}{Z_{T1}+Z_0}=\frac{Z_0/n^2-Z_0}{\frac{Z_0}{n^2}+Z_0}=\frac{1-n^2}{1+n^2}$$

$$S_{21}=2\sqrt{\frac{Z_{02}}{Z_{01}}}\frac{V_2}{E_1}=2\frac{V_2}{E_1}\quad(1)$$

$$V_1=\frac{E_1 Z_{T1}}{Z_{T1}+Z_0}=\frac{E_1\,Z_0/n^2}{\frac{Z_0}{n^2}+Z_0}=\frac{E_1}{n^2+1}$$

$$V_2=nV_1=\frac{nE_1}{n^2+1}\quad\text{OR}\quad\frac{V_2}{E_1}=\frac{n}{n^2+1}\quad(2)$$

(2) INTO (1):

$$S_{21}=\frac{2n}{n^2+1}$$

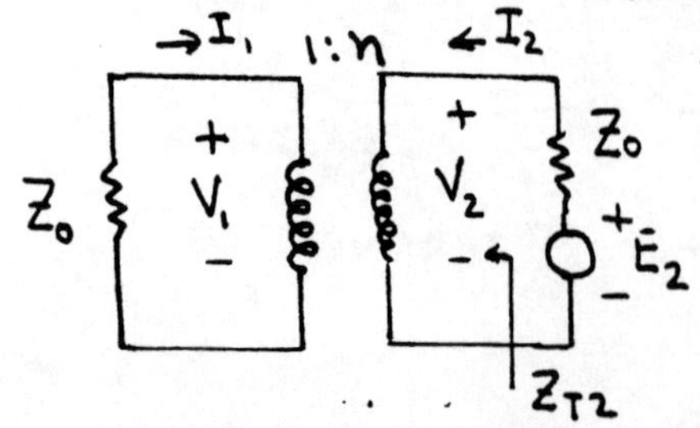

THE PARAMETERS S_{22} AND S_{12} ARE CALCULATED AS FOLLOWS:

$$Z_{T2}=Z_0 n^2$$

$$S_{22} = \left.\frac{b_2}{a_2}\right|_{a_1=0} = \frac{Z_{T2}-Z_0}{Z_{T2}+Z_0} = \frac{n^2 Z_0 - Z_0}{n^2 Z_0 + Z_0} = \frac{n^2-1}{n^2+1}$$

$$S_{12} = 2\frac{V_1}{E_2} \quad (3)$$

$$V_2 = \frac{E_2 Z_{T2}}{Z_{T2}+Z_0} = \frac{E_2 Z_0 n^2}{Z_0 n^2 + Z_0} = \frac{E_2 n^2}{n^2+1}$$

$$V_1 = \frac{V_2}{n} = \frac{E_2 n}{n^2+1} \quad \text{OR} \quad \frac{V_1}{E_2} = \frac{n}{n^2+1} \quad (4)$$

(4) INTO (3):

$$S_{12} = \frac{2n}{n^2+1}$$

AT PORTS 1' AND 2', WITH $\theta = \theta_1 = \theta_2$, WE OBTAIN:

$$[S'] = \begin{bmatrix} S_{11}e^{-j2\theta} & S_{12}e^{-j2\theta} \\ S_{21}e^{-j2\theta} & S_{22}e^{-j2\theta} \end{bmatrix} = e^{-j2\theta}\begin{bmatrix} \frac{1-n^2}{1+n^2} & \frac{2n}{n^2+1} \\ \frac{2n}{n^2+1} & \frac{n^2-1}{n^2+1} \end{bmatrix}$$

1.12)

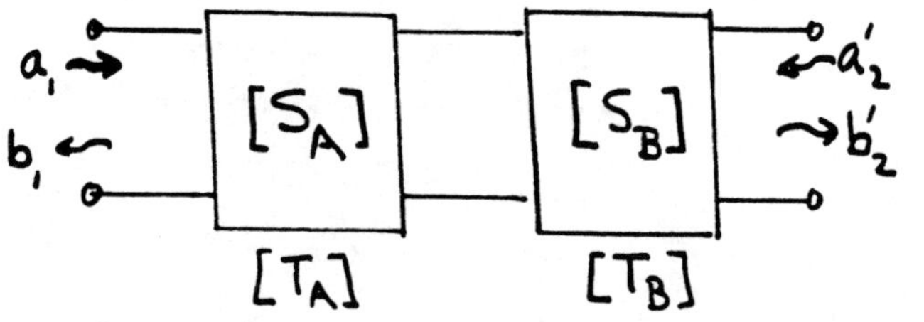

[T] AND [S] PARAMETERS ARE RELATED BY (1.4.11)

FROM (1.4.13):

$$\begin{bmatrix} a_1 \\ b_1 \end{bmatrix} = \begin{bmatrix} T_{11}^A & T_{12}^A \\ T_{21}^A & T_{22}^A \end{bmatrix}\begin{bmatrix} T_{11}^B & T_{12}^B \\ T_{21}^B & T_{22}^B \end{bmatrix}\begin{bmatrix} b_2' \\ a_2' \end{bmatrix}$$

OVERALL T_{11} IS: $T_{11} = T_{11}^A T_{11}^B + T_{12}^A T_{21}^B$

$$T_{11} = \frac{1}{S_{21}^A}\frac{1}{S_{21}^B} + \left(-\frac{S_{22}^A}{S_{21}^A}\right)\left(\frac{S_{11}^B}{S_{21}^B}\right)$$

SINCE $T_{11} = \frac{1}{S_{21}} = \frac{1-S_{22}^A S_{11}^B}{S_{21}^A S_{21}^B} \Rightarrow S_{21} = \frac{S_{21}^A S_{21}^B}{1-S_{22}^A S_{11}^B}$

1.13) (a) $\Gamma_0 = \dfrac{100-50}{100+50} = \dfrac{1}{3}$, $VSWR = \dfrac{1+\frac{1}{3}}{1-\frac{1}{3}} = 2$

(b) $Z_{IN}(\lambda/4) = \dfrac{Z_0^2}{Z_L} = \dfrac{50^2}{100} = 25\,\Omega$

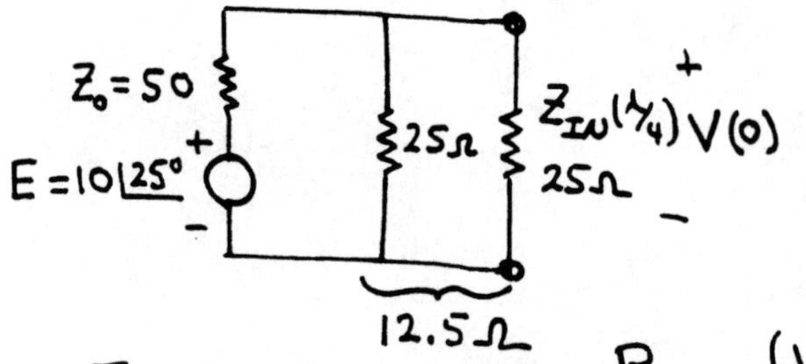

$V(0) = \dfrac{10\angle 25^\circ (12.5)}{12.5+50} = 2\angle 25^\circ$

$P_{AVS} = \dfrac{|E|^2}{8Z_0} = \dfrac{10^2}{8(50)} = 0.25\text{ W}$

THE INPUT POWER IS: $P_{IN} = \dfrac{(V(0)_{rms})^2}{12.5} = \dfrac{\left(\frac{2}{\sqrt{2}}\right)^2}{12.5} = 0.16\text{ W}$

SINCE THE TRANSMISSION LINE IS LOSSLESS, THE POWER DELIVERED TO THE LOAD (P_L) IS THE SAME AS P_{IN}.

$\therefore\ P_L = P_{IN} = 0.16\text{ W}$

1.14) (a)

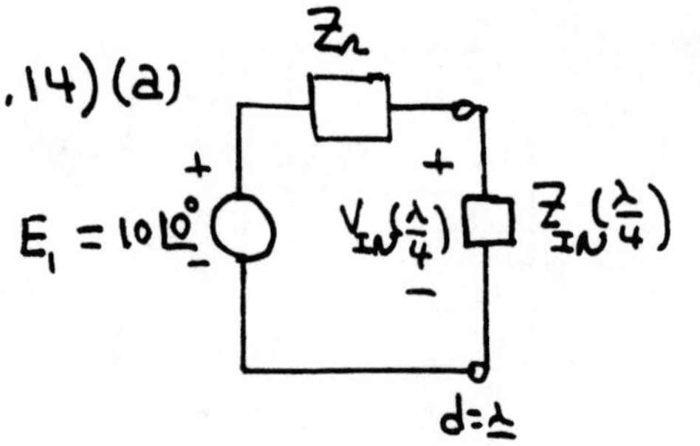

$Z_{IN}(\frac{\lambda}{4}) = \dfrac{Z_0^2}{Z_L} = \dfrac{50^2}{50+j50} = 25-j25$

NOTE: If Z_R IS GIVEN, THE VALUE OF Z_{IN} FOR MAXIMUM POWER IS $Z_{IN} = Z_R^*$. HOWEVER, IN THIS PROBLEM Z_{IN} IS GIVEN. HENCE, THE VALUE OF Z_R FOR MAX. POWER DELIVERED TO Z_{IN} IS: $Z_R = -\,\mathrm{Im}[Z_{IN}]$

$\therefore\ Z_R = j25$

$V_{IN}(\frac{\lambda}{4}) = \dfrac{10\angle 0^\circ (25-j25)}{j25+(25-j25)} = 10-j10 = 14.14\angle{-45^\circ}$

$I_{IN}(\frac{\lambda}{4}) = \dfrac{14.14\angle{-45^\circ}}{25-j25} = 0.4$

$P_{IN} = \frac{1}{2}\mathrm{Re}[14.14\angle{-45^\circ}(0.4)] = 2\text{ W}$

$P_L = P_{IN} = 2\text{ W}$

(b) TO FIND E_{TH}, WE FIND THE OPEN CIRCUIT VOLTAGE AT $d=0$:

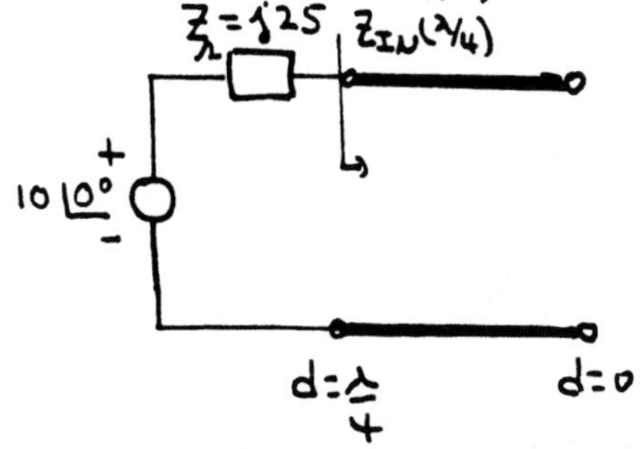

$$\Gamma_0 = 1, \quad V(d) = 2A\cos\beta d$$

$$I(d) = j\frac{2A}{Z_0}\sin\beta d \qquad (1)$$

$$Z_{IN}\left(\frac{\lambda}{4}\right) = 0 \quad \text{(A SHORT CIRCUIT)}$$

$$\therefore I\left(\frac{\lambda}{4}\right) = \frac{10\angle 0^\circ}{j25 + 0} = -j0.4 \qquad (2)$$

FROM (1) AND (2):

$$-j0.4 = j\frac{2A}{50}\sin\frac{\pi}{2} \Rightarrow A = -10$$

HENCE:

$$V(d) = 2(-10)\cos\beta d = -20\cos\beta d$$

$$E_{TH} = V(0) = -20$$

TO FIND Z_{TH} WE SET $E_1 = 0$, THEN:

$$Z_{TH} = \frac{Z_0^2}{Z_L} = \frac{50^2}{j25} = -j100$$

THE THEVENIN'S EQUIVALENT CIRCUIT AT $d=0$ IS:

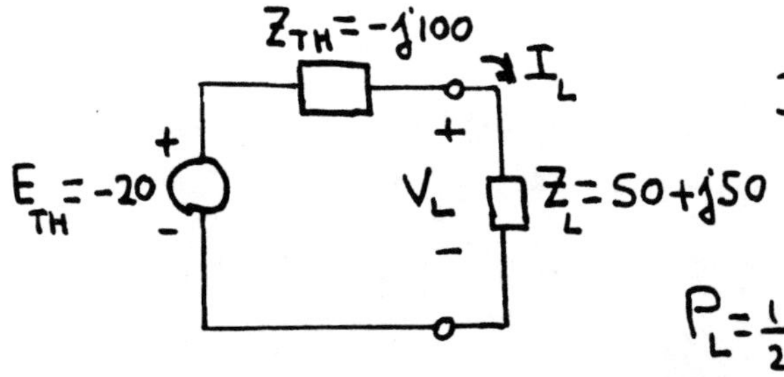

$$I_L = \frac{-20}{-j100 + 50 + j50} = 0.283\angle -135^\circ$$

$$V_L = 0.283\angle -135^\circ (50 + j50) = 20\angle -90^\circ$$

$$P_L = \frac{1}{2}\text{Re}\left[20\angle -90^\circ (0.283\angle 135^\circ)\right] = 2\,W$$

1.15)

$$v_2(0) = \frac{V_2(0)}{\sqrt{Z_0}} = a_2(0) + b_2(0) = 0 + 3.54\angle 45^\circ$$

$$\therefore V_2(0) = \sqrt{50}\,(3.54\angle 45^\circ) = 25.03\angle 45^\circ$$

$$i_2(0) = \sqrt{Z_0}\, I_2(0) = a_2(0) - b_2(0) = 0 - 3.54\angle 45^\circ$$

$$\therefore I_2(0) = \frac{-3.54\angle 45^\circ}{\sqrt{50}} = 0.501\angle -135^\circ$$

$$P_2(0) = \frac{1}{2}|I_2(0)|^2\, 50 = \frac{1}{2}(0.501)^2\, 50 = 6.27\,W$$

$$P_2(0) = \frac{1}{2}\frac{|V_2(0)|^2}{50} = \frac{1}{2}\frac{(25.03)^2}{50} = 6.27\,W$$

ALSO: $P_2(0) = \frac{1}{2}|a_2(0)|^2 - \frac{1}{2}|b_2(0)|^2 = \frac{1}{2}(3.54)^2 = 6.27\,W$

1.16) (a) $Z_{IN}(0) = Z_{IN}(d)\Big|_{d=\frac{\lambda}{8}} = 50\,\dfrac{(150+j150)+j50\tan 45^\circ}{50+j(150+j150)\tan 45^\circ} = 23 - j65\ \Omega$

(b) $a_1(0) = \dfrac{1}{2\sqrt{Z_0}}\left[V_1(0) + Z_0 I_1(0)\right]$, $V_1(0) = E_1 - Z_0 I_1(0)$

$$\therefore\ a_1(0) = \frac{E_1}{2\sqrt{Z_0}} = \frac{10}{2\sqrt{50}} = 0.707$$

$$a_1(\tfrac{\lambda}{8}) = a_1(0)\,e^{-j\pi/4} = 0.707\,\underline{|-45^\circ}$$

$$b_1(0) = \frac{1}{2\sqrt{50}}\left[V_1(0) - 50 I_1(0)\right] = \frac{1}{2\sqrt{50}}\left[10 - 50 I_1(0) - 50 I_1(0)\right] \quad (1)$$

$$I_1(0) = \frac{10\,\underline{|0^\circ}}{50+23-j65} = 0.102\,\underline{|41.68^\circ} \quad (2)$$

(2) INTO (1): $b_1(0) = \dfrac{1}{2\sqrt{50}}\left[10 - 2(50)(0.102\,\underline{|41.68^\circ})\right] = 0.508\,\underline{|-70.65^\circ}$

$$b_1(\tfrac{\lambda}{8}) = b_1(0)\,e^{j\pi/4} = 0.508\,\underline{|-25.65^\circ}$$

SINCE $Z_2 = Z_0$, THE OUTPUT IS MATCHED. HENCE, $a_2(0) = 0$

(c) $V_1(0) = I_1(0)\,Z_{IN}(0) = 0.102\,\underline{|41.68^\circ}\,(23-j65) = 7.05\,\underline{|-28.8^\circ}$

OR $V_1(0) = \sqrt{Z_0}\left[a_1(0) + b_1(0)\right] = \sqrt{50}\left[0.707 + 0.508\,\underline{|-70.65^\circ}\right] = 7.05\,\underline{|-28.8^\circ}$

$V_1(\tfrac{\lambda}{8}) = \sqrt{Z_0}\left[a_1(\tfrac{\lambda}{8}) + b_1(\tfrac{\lambda}{8})\right] = \sqrt{50}\left[0.707\,\underline{|-45^\circ} + 0.508\,\underline{|-25.65^\circ}\right] = 8.47\,\underline{|-36.92^\circ}$

$I_1(\tfrac{\lambda}{8}) = \dfrac{8.47\,\underline{|-36.92^\circ}}{150+j150} = 0.04\,\underline{|-81.92^\circ}$

(d),(e) $P_1(0) = \frac{1}{2}\mathrm{Re}\left[V_1(0) I_1^*(0)\right] = 0.12\,W$, $P_1(\tfrac{\lambda}{8}) = \frac{1}{2}\mathrm{Re}\left[V_1(\tfrac{\lambda}{8}) I_1^*(\tfrac{\lambda}{8})\right] = 0.12\,W$

ALSO: $P_1(0) = P_1(\tfrac{\lambda}{8}) = \frac{1}{2}|a_1(0)|^2 - \frac{1}{2}|b_1(0)|^2 = \frac{1}{2}|a_1(\tfrac{\lambda}{8})|^2 - \frac{1}{2}|b_1(\tfrac{\lambda}{8})|^2 = \frac{1}{2}(0.707)^2 - \frac{1}{2}(0.508)^2 = 0.12\,W$

(f) $S_{11}(\tfrac{\lambda}{8}) = \dfrac{Z_{T1} - Z_0}{Z_{T1} + Z_0} = \dfrac{150+j150-50}{150+j150+50} = 0.721\,\underline{|19.44^\circ}$

$S_{11}(0) = S_{11}(\tfrac{\lambda}{8})\,e^{-j2(\pi/4)} = 0.721\,\underline{|19.44^\circ}\,(\underline{|-90^\circ}) = 0.721\,\underline{|-70.56^\circ}$

(g) $(VSWR)_{in} = \dfrac{1+|S_{11}|}{1-|S_{11}|} = 6.17$, $(VSWR)_{out} = 1$

(h) $\lambda = \dfrac{3\,10^8}{10^9} = 0.3\,m$ (OR 30 cm) . $l = \dfrac{\lambda}{8} = \dfrac{30}{8} = 3.75$ cm

(i) $b_2(\tfrac{\lambda}{8}) = S_{21}\,a_1(\tfrac{\lambda}{8}) + S_{22}\,a_2(\tfrac{\lambda}{8})^{\,0} = 3\,\underline{|60^\circ}\,(0.707\,\underline{|-45^\circ}) = 2.12\,\underline{|15^\circ}$

$$P_2(0) = \tfrac{1}{2}|b_2(0)|^2 = \tfrac{1}{2}|b_2(\tfrac{\lambda}{8})|^2 = \tfrac{1}{2}(2.12)^2 = 2.25\ W$$

1.17) (a) $Z_{IN}(0) = Z_{IN}(d)\Big|_{d=\frac{\lambda}{4}} = 75\,\dfrac{(150+j150)+j75\tan 45^\circ}{75+j(150+j150)\tan 45^\circ} = 60 - j105\ \Omega$

(b) THE VSWR IN THE $l_2 = \frac{\lambda}{4}$ LINE IS UNITY, SINCE THE LINE IS MATCHED.

IN THE $l_1 = \frac{\lambda}{8}$ LINE : $\Gamma_0 = \dfrac{(150+j150)-75}{(150+j150)+75} = 0.62\angle 29.7^\circ$

$\therefore$ VSWR $= \dfrac{1+0.62}{1-0.62} = 4.26$

(c) $V_1(0) = \dfrac{10\angle 0^\circ\, Z_{IN}(0)}{Z_{IN}(0)+Z_1} = \dfrac{10(60-j105)}{60-j105+100} = 6.32\angle -26.98^\circ$

$I_1(0) = \dfrac{V_1(0)}{Z_{IN}(0)} = \dfrac{6.32\angle -26.98^\circ}{60-j105} = 0.0523\angle 33.27^\circ$

$a_1(0) = \dfrac{1}{2\sqrt{75}}\left[V_1(0) + 75\,I_1(0)\right] = 0.516\angle -4.6^\circ$

$b_1(0) = \dfrac{1}{2\sqrt{75}}\left[V_1(0) - 75\,I_1(0)\right] = 0.32\angle -64.88^\circ$

$a_1(\frac{\lambda}{8}) = a_1(0)e^{-j\pi/4} = 0.516\angle -49.6^\circ$

$b_1(\frac{\lambda}{8}) = b_1(0)e^{j\pi/4} = 0.32\angle -19.8^\circ$

$a_2(0) = 0$

(d) $P_1(0) = \frac{1}{2}|a_1(0)|^2 - \frac{1}{2}|b_1(0)|^2 = \frac{1}{2}(0.516)^2 - \frac{1}{2}(0.32)^2 = 0.082\ W$

$P_1(\frac{\lambda}{8}) = \frac{1}{2}|a_1(\frac{\lambda}{8})|^2 - \frac{1}{2}|b_1(\frac{\lambda}{8})|^2 = \frac{1}{2}(0.516)^2 - \frac{1}{2}(0.32)^2 = 0.082\ W$

(e) $P_{AVS} = \dfrac{|E_1|^2}{8\,\mathrm{Re}[Z_1]} = \dfrac{(10)^2}{8(100)} = 0.125\ W$

$\frac{1}{2}|a_1(0)|^2 = 0.133\ W$

SINCE $Z_1 \neq Z_0$, IT FOLLOWS THAT $P_{AVS} \neq \frac{1}{2}|a_1(0)|^2$

1.18) (a) $V = \dfrac{5\angle 30^\circ(100)}{100+(50+j50)} = 3.162\angle 11.565^\circ$

$I = \dfrac{V}{Z_L} = \dfrac{3.162\angle 11.565^\circ}{100} = 0.0316\angle 11.565^\circ$

$a_p = \dfrac{1}{2\sqrt{R_\Omega}}(V + Z_\Omega I) = \dfrac{1}{2\sqrt{50}}\left(3.162\angle 11.565^\circ + (50+j50)(0.0316\angle 11.565^\circ)\right)$

$= 0.355\angle 30.14^\circ$

$$b_p = \frac{1}{2\sqrt{R_r}}(V - Z_r^* I) = \frac{1}{2\sqrt{50}}\left(3.162\angle 11.565° - (50-j50)(0.0316\angle 11.565°)\right)$$

$$= 0.158\angle 56.53°$$

$$V_p^+ = \frac{E_r Z_r^*}{2R_r} = \frac{5\angle 30°(50-j50)}{2(50)} = 3.536\angle -15°$$

$$V_p^- = V - V_p^+ = 3.162\angle 11.565° - 3.536\angle -15° = 1.581\angle 101.59°$$

$$I_p^+ = \frac{E_r}{2R_r} = \frac{5\angle 30°}{2(50)} = 0.05\angle 30°$$

$$I_p^- = I_p^+ - I = 0.05\angle 30° - 0.0316\angle 11.565° = 0.022\angle 56.52°$$

(b)
(c) $$V_p^+ = \frac{Z_r^*}{\sqrt{R_r}} a_p = \frac{(50-j50)}{\sqrt{50}} 0.355\angle 30.14° = 3.54\angle -15°$$

$$V_p^- = \frac{Z_r}{\sqrt{R_r}} b_p = \frac{(50+j50)}{\sqrt{50}} 0.158\angle 56.53° = 1.58\angle 101.5°$$

$$V = V_p^+ + V_p^- = 3.16\angle 11.56°$$

$$I_p^+ = \frac{a_p}{\sqrt{R_r}} = \frac{0.355\angle 30.14°}{\sqrt{50}} = 0.05\angle 30.14°$$

$$I_p^- = \frac{b_p}{\sqrt{R_r}} = \frac{0.158\angle 56.53°}{\sqrt{50}} = 0.022\angle 56.53°$$

$$I = I_p^+ - I_p^- = 0.0316\angle 11.56°$$

1.19) IN EXAMPLE 1.71: $V = 5.59\angle -26.57°$, $a_p = 0.5$, $b_p = 0$

(a) $$V_p^+ = \frac{E_r Z_r^*}{2R_r} = \frac{10(100-j50)}{2(100)} = 5.59\angle -26.57°$$

$$V_p^- = V - V_p^+ = 5.59\angle -26.57° - 5.59\angle -26.57° = 0$$

(b) $$a_p = \frac{\sqrt{R_r}}{Z_r^*} V_p^+ = \frac{\sqrt{100}}{100-j50}(5.59\angle -26.57°) = 0.5$$

$$b_p = \frac{\sqrt{R_r}}{Z_r} V_p^- = 0$$

1.20) IN EXAMPLE 1.7.2: $V = 10$, $a_p = 1.5$, $b_p = -0.5$

(a) $V_p^+ = \dfrac{E_\Omega Z_\Omega^*}{2R_\Omega} = \dfrac{30\,(100)}{2(100)} = 15$

$V_p^- = V - V_p^+ = 10 - 15 = -5$

(b) $a_p = \dfrac{\sqrt{R_\Omega}}{Z_\Omega^*} V_p^+ = \dfrac{\sqrt{100}}{100}(15) = 1.5$

$b_p = \dfrac{\sqrt{R_\Omega}}{Z_\Omega} V_p^- = \dfrac{\sqrt{100}}{100}(-5) = -0.5$

1.21) WITH $E_2 = 0$: $Z_{T1} = 100 + 200 = 300$

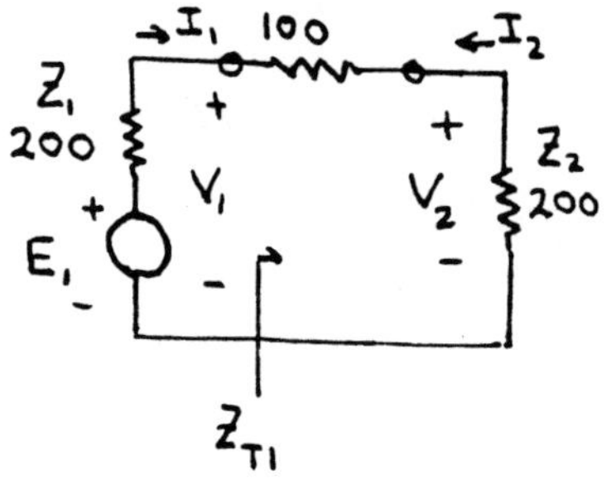

$S_{p11} = \dfrac{Z_{T1} - Z_1^*}{Z_{T1} + Z_1} = \dfrac{300-200}{300+200} = 0.2$

$V_1 = \dfrac{E_1\, 300}{300+200} = \dfrac{3}{5}E_1$, $I_1 = \dfrac{E_1}{500}$

$V_2 = \dfrac{E_1\, 200}{200+300} = \dfrac{2}{5}E_1$, $I_2 = \dfrac{-E_1}{500}$

$a_{p1} = \dfrac{1}{2\sqrt{200}}(V_1 + Z_1 I_1) = \dfrac{1}{2\sqrt{200}}\left(\dfrac{3}{5}E_1 + 200\dfrac{E_1}{500}\right) = \dfrac{E_1}{2\sqrt{200}}$, $a_{p2} = 0$

$b_{p2} = \dfrac{1}{2\sqrt{200}}(V_2 - Z_2^* I_2) = \dfrac{1}{2\sqrt{200}}\left(\dfrac{2}{5}E_1 + 200\dfrac{E_1}{500}\right) = \dfrac{2E_1}{5\sqrt{200}}$

$S_{p21} = \dfrac{b_{p2}}{a_{p1}}\Bigg|_{a_{p2}=0} = \dfrac{2E_1\; 2\sqrt{200}}{5\sqrt{200}\; E_1} = \dfrac{4}{5}$

FROM SYMMETRY: $S_{p22} = S_{p11}$ AND $S_{p12} = S_{p21}$

THE MAGNITUDE OF S_{p21} CAN ALSO BE CALCULATED AS FOLLOWS:

$|S_{p21}|^2 = \dfrac{P_L}{P_{AVS}}$, $P_L = \dfrac{1}{2}\dfrac{|E_1(\frac{2}{5})|^2}{200}$, $P_{AVS} = \dfrac{|E_1|^2}{8(200)}$

$\therefore |S_{p21}|^2 = \dfrac{16}{25}$ OR $|S_{p21}| = \dfrac{4}{5}$

SINCE IN THIS EXAMPLE S_{p21} IS REAL, THEN $S_{p21} = \dfrac{4}{5}$.

1.22) (a) $f = 1\,\text{GHz}$

EVALUATION OF S_{p11} AND S_{p21} (LET $E_1 = 1\angle 0^\circ$)

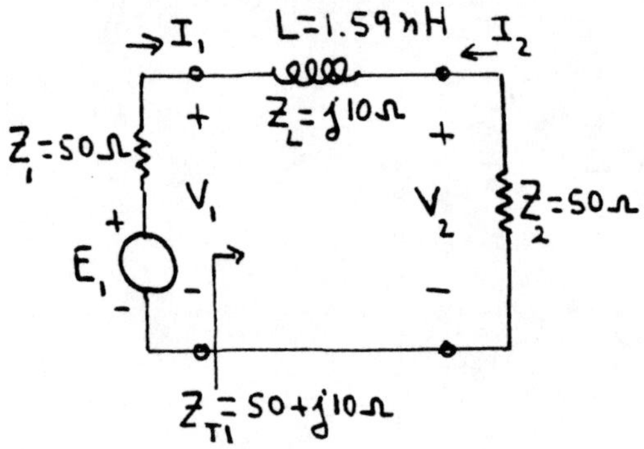

$$V_1 = \frac{E_1(50+j10)}{100+j10} = 0.5074\angle 5.599^\circ$$

$$I_1 = \frac{V_1}{50+j10} = 0.01\angle -5.711^\circ$$

$$V_2 = \frac{E_1\,50}{100+j10} = 0.4975\angle -5.711^\circ$$

$$I_2 = -\frac{V_2}{50} = 0.01\angle 174.29^\circ$$

$$a_{p1} = \frac{1}{2\sqrt{R_1}}(V_1 + Z_1 I_1) = \frac{1}{2\sqrt{50}}\left(0.5074\angle 5.599^\circ + 50(0.01\angle -5.711^\circ)\right) = 0.0709\angle -0.0143^\circ$$

$$b_{p1} = \frac{1}{2\sqrt{R_1}}(V_1 - Z_1^* I_1) = \frac{1}{2\sqrt{50}}\left(0.5074\angle 5.599^\circ - 50(0.01\angle -5.711^\circ)\right) = 0.007\angle 85.70^\circ$$

$$a_{p2} = \frac{1}{2\sqrt{R_2}}(V_2 + Z_2 I_2) = 0$$

$$b_{p2} = \frac{1}{2\sqrt{R_2}}(V_2 - Z_2^* I_2) = \frac{1}{2\sqrt{50}}\left(0.4975\angle -5.711^\circ - 50(0.01\angle 174.29^\circ)\right) = 0.0705\angle -5.711^\circ$$

$$S_{p11} = \left.\frac{b_{p1}}{a_{p1}}\right|_{a_{p2}=0} = \frac{0.007\angle 85.7^\circ}{0.0709\angle -0.0143^\circ} = 0.099\angle 85.7^\circ$$

S_{p11} CAN ALSO BE CALCULATED USING:

$$S_{p11} = \frac{Z_{T1} - Z_1^*}{Z_{T1} + Z_1} = \frac{50+j10-50}{50+j10+50} = 0.099\angle 84.3^\circ$$

$$S_{p21} = \left.\frac{b_{p2}}{a_{p1}}\right|_{a_{p2}=0} = \frac{0.0705\angle -5.711^\circ}{0.0709\angle -0.0143^\circ} = 0.994\angle -5.7^\circ$$

(b) WITH $Z_0 = 50\,\Omega$, IT FOLLOWS FROM (1.6.24) AND (1.6.25) THAT

$$S_{11} = S_{22} = \frac{Z_L}{Z_L + 2Z_0} = \frac{j10}{j10+100} = 0.099\angle 84.3^\circ$$

$$S_{21} = S_{12} = \frac{2Z_0}{Z_L + 2Z_0} = \frac{100}{j10+100} = 0.995\angle -5.7^\circ$$

AS EXPECTED, FOR $Z_1 = Z_2 = Z_0 = 50\,\Omega$ THE S_p PARAMETERS ARE IDENTICAL TO THE S PARAMETERS.

1.23) AT P: $I_1^+ = I_1^- + I_2^- + I_3^-$

AND $b_1 = S_{11} a_1 \Rightarrow \sqrt{Z_0}\, I_1^- = S_{11}\sqrt{Z_0}\, I_1^+$ OR $I_1^- = S_{11} I_1^+$

$b_2 = S_{21} a_1 \Rightarrow \sqrt{Z_0}\, I_2^- = S_{21}\sqrt{Z_0}\, I_1^+$ OR $I_2^- = S_{21} I_1^+$

$b_3 = S_{31} a_1 \Rightarrow \sqrt{Z_0}\, I_3^- = S_{31}\sqrt{Z_0}\, I_1^+$ OR $I_3^- = S_{31} I_1^+$

$$\therefore \quad I_1^+ = S_{11} I_1^+ + S_{21} I_1^+ + S_{31} I_1^+$$

$$S_{11} + S_{21} + S_{31} = 1$$

SIMILARLY, WITH E_2 APPLIED TO PORT 2 WITH THE OTHER PORTS MATCHED GIVES

$$S_{12} + S_{22} + S_{32} = 0$$

WITH E_3 APPLIED TO PORT 3, WITH THE OTHER PORTS MATCHED GIVES

$$S_{13} + S_{23} + S_{33} = 0$$

WITH ALL THE SOURCES EQUAL TO E_0, AS SHOWN IN FIG. 1.22b, THE CURRENTS I_1, I_2, AND I_3 ARE EQUAL TO ZERO. HENCE,

$I_1^+ = I_1^-$, $I_2^+ = I_2^-$, $I_3^+ = I_3^-$, AND $I_1^+ = I_2^+ = I_3^+$

FROM

$$b_1 = S_{11} a_1 + S_{12} a_2 + S_{13} a_3$$

OR

$$I_1^- = S_{11} I_1^+ + S_{12} I_2^+ + S_{13} I_3^+ \Rightarrow S_{11} + S_{12} + S_{13} = 1$$

SIMILARLY:

$$I_2^- = S_{21} I_1^+ + S_{22} I_2^+ + S_{23} I_3^+ \Rightarrow S_{21} + S_{22} + S_{23} = 1$$

$$I_3^- = S_{31} I_1^+ + S_{32} I_2^+ + S_{33} I_3^+ \Rightarrow S_{31} + S_{32} + S_{33} = 1$$

1.24) (a)

$v_1 = z_{11} i_1 + z_{12} i_2$ (1) $\qquad$ $i_1 = y_{11} v_1 + y_{12} v_2$ (3)

$v_2 = z_{21} i_1 + z_{22} i_2$ (2) $\qquad$ $i_2 = y_{21} v_1 + y_{22} v_2$ (4)

FROM (2): $i_2 = \dfrac{v_2}{z_{22}} - \dfrac{z_{21}}{z_{22}} i_1$ (5)

(5) INTO (1): $i_1 = \dfrac{z_{22}}{|z|} v_1 - \dfrac{z_{12}}{|z|} v_2$ (6), $|z| = z_{11} z_{22} - z_{21} z_{12}$

COMPARING (6) WITH (3) GIVES: $y_{11} = \dfrac{z_{22}}{|z|}$ AND $y_{12} = -\dfrac{z_{12}}{|z|}$

(6) INTO (2): $v_2 = z_{21}\left(\frac{z_{22}}{|z|} v_1 - \frac{z_{12}}{|z|} v_2\right) + z_{22} i_2$

OR $i_2 = -\frac{z_{21}}{|z|} v_1 + \frac{z_{11}}{|z|} v_2$ (7)

COMPARING (7) WITH (4) GIVES: $y_{21} = -\frac{z_{21}}{|z|}$ AND $y_{22} = \frac{z_{11}}{|z|}$.

THE DERIVATION BETWEEN y AND z PARAMETERS IS SIMILAR.

(b) $v_1 = A v_2 - B i_2$ (8)

$i_1 = C v_2 - D i_2$ (9)

FROM (9): $v_2 = \frac{1}{C} i_1 + \frac{D}{C} i_2$ (10)

(10) INTO (8): $v_1 = \frac{A}{C} i_1 + \frac{AD-BC}{C} i_2$ (11)

COMPARING (11) WITH (1) GIVES: $z_{11} = \frac{A}{C}$ AND $z_{12} = \frac{AD-BC}{C}$

(11) INTO (8): $\frac{A}{C} i_1 + \frac{AD-BC}{C} i_2 = A v_2 - B i_2$

OR $v_2 = \frac{1}{C} i_1 + \frac{D}{C} i_2$ (12)

COMPARING (12) WITH (2) GIVES: $z_{21} = \frac{1}{C}$ AND $z_{22} = \frac{D}{C}$

THE DERIVATION BETWEEN z AND ABCD IS SIMILAR.

1.25)(a) FOR REAL $[Z_0]$ AND $[Y_0]$

$$[I] = [y][V]$$

$$[I^-] = [I^+] - [I]$$

$$[V^-][Y_0] = [V^+][Y_0] - [y]([V^+] + [V^-])$$

$$([Y_0] + [y])[V^-] = ([Y_0] - [y])[V^+]$$

$$[V^-] = ([Y_0] + [y])^{-1}([Y_0] - [y])[V^+]$$

$$= [S][V^+]$$

$$\therefore [S] = -([Y_0] + [y])^{-1}([y] - [Y_0]) \quad (1)$$

$$[Y_o] - [y] = [Y_o][S] + [y][S]$$

$$[Y_o]([I] - [S]) = [y]([I] + [S])$$

$$[y] = [Y_o]([I] - [S])([I] + [S])^{-1}$$

(b) FROM (1):

$$[S] = -\left[\begin{bmatrix} Y_o & 0 \\ 0 & Y_o \end{bmatrix} + \begin{bmatrix} y_{11} & y_{12} \\ y_{21} & y_{22} \end{bmatrix}\right]^{-1} \left[\begin{bmatrix} y_{11} & y_{12} \\ y_{21} & y_{22} \end{bmatrix} - \begin{bmatrix} Y_o & 0 \\ 0 & Y_o \end{bmatrix}\right]$$

$$[S] = -\begin{bmatrix} y_{11}+Y_o & y_{12} \\ y_{21} & y_{22}+Y_o \end{bmatrix}^{-1} \begin{bmatrix} y_{11}-Y_o & y_{12} \\ y_{21} & y_{22}-Y_o \end{bmatrix}$$

$$\begin{bmatrix} y_{11}+Y_o & y_{12} \\ y_{21} & y_{22}+Y_o \end{bmatrix}^{-1} = \frac{\begin{bmatrix} y_{22}+Y_o & -y_{12} \\ -y_{21} & y_{11}+Y_o \end{bmatrix}}{(y_{11}+Y_o)(y_{22}+Y_o) - y_{12}y_{21}}$$

$$\therefore \quad [S] = -\frac{\begin{bmatrix} (y_{22}+Y_o)(y_{11}-Y_o) - y_{12}y_{21} & (y_{22}+Y_o)y_{12} - y_{12}(y_{22}-Y_o) \\ -y_{21}(y_{11}-Y_o) + y_{21}(y_{11}+Y_o) & -y_{12}y_{21} + (y_{11}+Y_o)(y_{22}-Y_o) \end{bmatrix}}{(y_{11}+Y_o)(y_{22}+Y_o) - y_{12}y_{21}}$$

$$[S] = \frac{1}{\Delta_2}\begin{bmatrix} (1-y'_{11})(1+y'_{22}) + y'_{12}y'_{21} & -2y'_{12} \\ -2y'_{21} & (1+y'_{11})(1-y'_{22}) + y'_{12}y'_{21} \end{bmatrix}$$

WHERE $\Delta_2 = (1+y'_{11})(1+y'_{22}) - y'_{12}y'_{21}$, $y'_{11} = y_{11}Z_0$, $y'_{12} = y_{12}Z_0$, $y'_{21} = y_{21}Z_0$, AND $y'_{22} = y_{22}Z_0$ ($Y_o = 1/Z_0$).

1.26) STEP 1. CONVERT THE [S] PARAMETERS OF THE BJT TO [z] PARAMETERS (CALLED $[z_1]$).

STEP 2. CALCULATE THE [z] PARAMETERS OF THE INDUCTOR (CALLED $[z_2]$).

STEP 3. ADD THE [z] PARAMETERS (i.e., $[z_3] = [z_1] + [z_2]$)

STEP 4. CONVERT $[z_3]$ TO [S] PARAMETERS. THESE ARE THE [S] PARAMETERS OF THE TWO-PORT NETWORK.

1.27) CONVERT THE CE S PARAMETERS TO CE y PARAMETERS (SEE FIG. 1.8.1), THAT IS:

$$y_{11,e}=0.016+j0.034 \qquad y_{12,e}=0.00141-j0.0272(10^{-3})$$
$$y_{21,e}=0.04-j0.036 \qquad y_{22,e}=0.00363+j0.0079$$

USE THE RELATIONS IN FIG. 1.8.1b TO CALCULATE THE CB AND CC y PARAMETERS. THAT IS,

$$y_{11,b}=0.061+j0.00587 \qquad y_{12,b}=-0.00503-j0.00787$$
$$y_{21,b}=-0.0436+j0.0281 \qquad y_{22,b}=0.00363+j0.0079$$

AND

$$y_{11,c}=0.016+j0.034 \qquad y_{12,c}=-0.0174-j0.034$$
$$y_{21,c}=-0.056+j0.002 \qquad y_{22,c}=0.061+j0.00587$$

CONVERT FROM CB AND CC y PARAMETERS TO CB AND CC S PARAMETERS. THAT IS,

$$[S]_{CB}=\begin{bmatrix}0.356\angle{-173.6^\circ} & 0.243\angle{35.41^\circ}\\ 1.348\angle{-54.81^\circ} & 1.198\angle{-32.7^\circ}\end{bmatrix} \text{ AND } [S]_{CC}=\begin{bmatrix}0.893\angle{-62.03^\circ} & 0.764\angle{29.68^\circ}\\ 1.12\angle{-35.26^\circ} & 0.176\angle{98.35^\circ}\end{bmatrix}$$

1.28) (a)

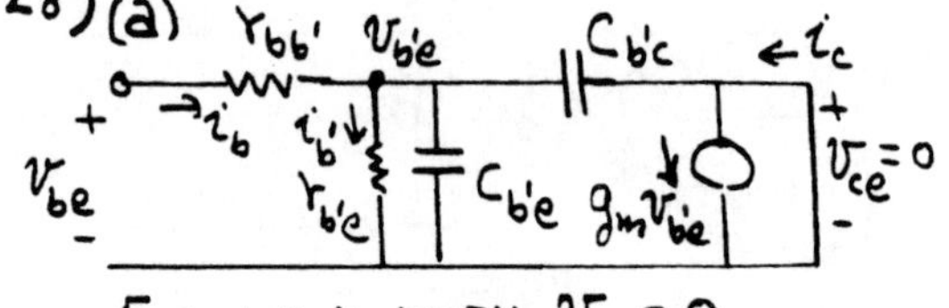

FIG. 1.11.4 WITH $v_{ce}=0$

$$i_c=g_m v_{b'e} \quad (1)$$
$$i_{b'}=i_b\frac{\left(\frac{1}{j\omega(C_{b'e}+C_{b'c})}\right)}{r_{b'e}+\left(\frac{1}{j\omega(C_{b'e}+C_{b'c})}\right)} \quad (2)$$
$$v_{b'e}=i_{b'}r_{b'e} \quad (3)$$

FROM (1), (2), AND (3): $i_c=g_m r_{b'e}i_{b'}=\dfrac{g_m r_{b'e}i_b}{1+j\omega r_{b'e}(C_{b'e}+C_{b'c})}$

DEFINE $h_{fe}(j\omega)$ AS:

$$h_{fe}(j\omega)=\frac{i_c}{i_b}\bigg|_{v_{ce}=0}=\frac{g_m r_{b'e}}{1+j\omega r_{b'e}(C_{b'e}+C_{b'c})}=\frac{g_m r_{b'e}}{1+j\omega/\omega_\beta}$$

WHERE

$$f_\beta=\frac{\omega_\beta}{2\pi}=\frac{1}{2\pi r_{b'e}(C_{b'e}+C_{b'c})}\approx\frac{1}{2\pi r_{b'e}C_{b'e}} \quad (\text{SINCE } C_{b'e}\gg C_{b'c})$$

f_T (THE GAIN-BANDWIDTH PRODUCT) IS THE FREQUENCY WHERE $|h_{fe}(j\omega)|=1$, OR

$$1=\frac{g_m r_{b'e}}{\sqrt{1+(\omega_T/\omega_\beta)^2}}\approx g_m r_{b'e}\frac{\omega_\beta}{\omega_T}\Rightarrow f_T=g_m r_{b'e}f_\beta=\frac{g_m}{2\pi C_{b'e}}$$

(b) SIMILARLY, FOR THE FET IN FIG. 1.11.15 WE OBTAIN

$$f_T=\frac{g_m}{2\pi(C_{gs}+C_{gd})}\approx\frac{g_m}{2\pi C_{gs}}=\frac{g_m}{2\pi C_i} \quad (\text{SINCE } C_{gs} \text{ IS DENOTED BY } C_i \text{ IN FIG. 1.11.15})$$

2.1) (a) LET $z_1 = \frac{Z}{50} = \frac{100 + j100}{50} = 2 + j2$

(z_1 IS SHOWN IN THE SMITH CHART)

(b) $y_1 = \frac{1}{z_1} = 0.25 - j0.25$

(y_1 IS SHOWN IN THE SMITH CHART)

(c) IN THE ZY SMITH CHART THE LOCATION OF z_1 IS THE SAME AS IN (a). THE VALUE OF y_1 IS READ FROM THE GREEN CIRCLES.

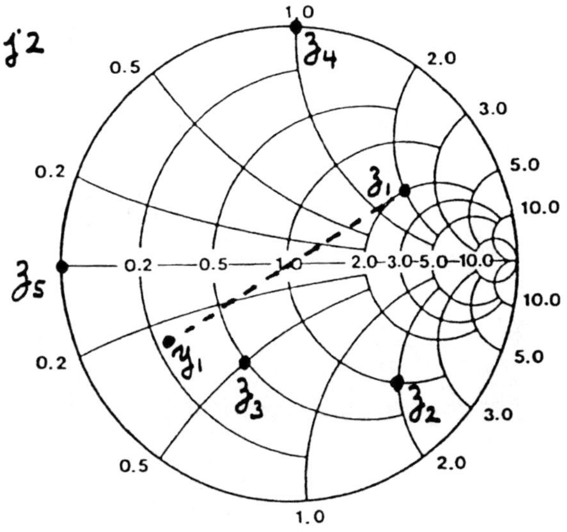

(d) $z_2 = \frac{50 - j100}{50} = 1 - j2$

$y_2 = \frac{1}{z_2} = 0.2 + j0.4$

$z_3 = \frac{25 - j25}{50} = 0.5 - j0.5$ $z_4 = j\frac{50}{50} = j1$ $z_5 = j0$

$y_3 = \frac{1}{z_3} = 1 + j$ $y_4 = \frac{1}{z_4} = -j1$ $y_5 = \infty$

(z_2, z_3, z_4, AND z_5 ARE SHOWN IN THE SMITH CHART)

2.2) (a) $z = -r + jx$, $\Gamma = \frac{z-1}{z+1} = \frac{-(r+1) + jx}{(1-r) + jx}$ (1)

$$|\Gamma| = \frac{\sqrt{(r+1)^2 + x^2}}{\sqrt{(1-r)^2 + x^2}} = \left[\frac{r^2 + 2r + 1 + x^2}{r^2 - 2r + 1 + x^2}\right]^{1/2} \quad (2)$$

IN (1) THE NUMERATOR IS GREATER THAN THE DENOMINATOR, SINCE $2r > -2r$. HENCE, $|\Gamma| > 1$

(b) FROM (1): $\frac{1}{\Gamma^*} = \frac{(1-r) - jx}{-(r+1) - jx} = \frac{(r-1) + jx}{(r+1) + jx}$ (3)

EQ. (3) IS IDENTICAL TO THE TRANSFORMATION IN (2.2.2) WHERE $z = r + jx$ WITH $r \geq 0$. HENCE, NEGATIVE RESISTANCES CAN BE HANDLED IN THE SMITH CHART BY PLOTTING $\frac{1}{\Gamma^*}$ AND INTERPRETING THE RESISTANCE CIRCLES AS BEING NEGATIVE, AND THE REACTANCE CIRCLES AS MARKED.

(c) $z_1 = \frac{Z_1}{50} = \frac{-20+j16}{50} = -0.4+j0.32$

FROM THE SMITH CHART WE READ:

$$\frac{1}{\Gamma^*} = 0.47\angle 139°$$

HENCE $\Gamma = 2.1\angle 139°$

FOR:

$$z_2 = \frac{-200+j25}{50} = -4+j0.5$$

$$\frac{1}{\Gamma^*} = 0.61\angle 3.75°$$

HENCE $\Gamma = 1.6\angle 3.75°$

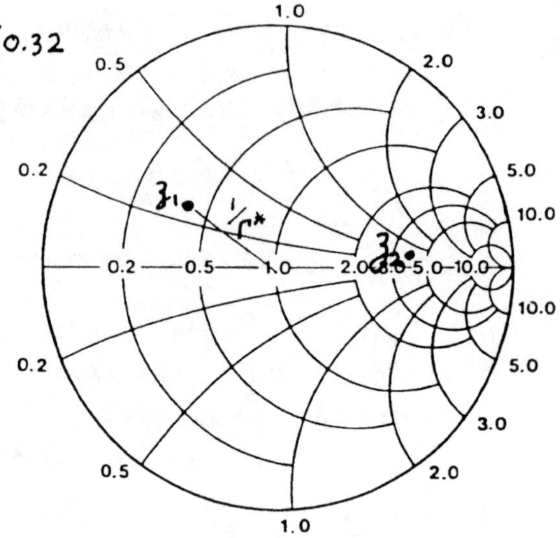

(d) SEE THE COMPRESSED SMITH CHART. FOR $z_1 = -0.4+j0.32$ WE READ: $|\Gamma| = 2.1$, $\angle\Gamma = 139°$. HENCE $\Gamma = 2.1\angle 139°$.

SIMILARLY, FOR z_2 WE OBTAIN $\Gamma = 1.6\angle 3.75°$

(e) $|\Gamma| \approx 3.2$

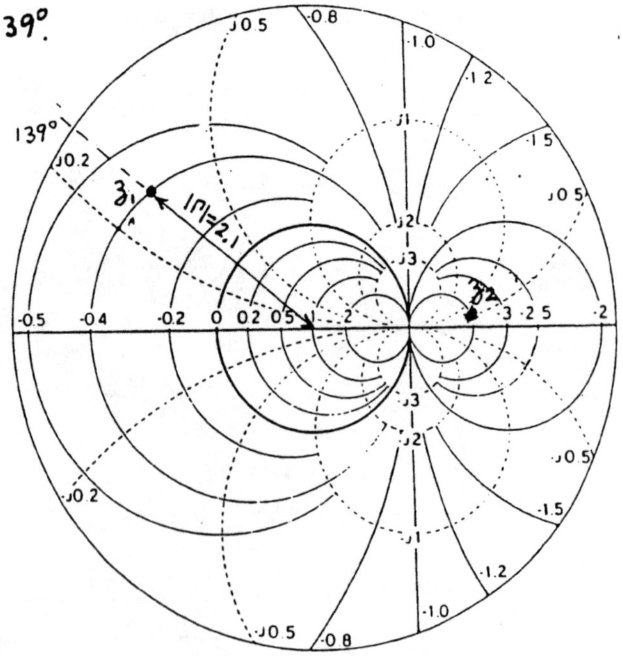

2.3) $Z(d) = Z_0 \frac{Z_L + jZ_0 \tan\beta d}{Z_0 + jZ_L \tan\beta d}$

$$Z\left(d+\frac{n\lambda}{2}\right) = Z_0 \frac{Z_L + jZ_0 \tan\left(\beta d + \frac{n\beta\lambda}{2}\right)}{Z_0 + jZ_L \tan\left(\beta d + \frac{n\beta\lambda}{2}\right)} \quad ; \; n\beta\frac{\lambda}{2} = n\frac{2\pi}{\lambda}\frac{\lambda}{2} = n\pi$$

$$\tan(\beta d + n\pi) = \tan\beta d$$

HENCE,

$$Z\left(d+\frac{n\lambda}{2}\right) = Z(d)$$

2.4) FROM (2.2.6) AND (2.2.7): LET $\Gamma_0 = |\Gamma_0| e^{j\phi_\ell}$, THEN

$$\Gamma = \Gamma_0 e^{-j2\beta d} = |\Gamma_0| e^{j(\phi_\ell - 2\beta d)} = |\Gamma| e^{j\phi}; \quad |\Gamma| = |\Gamma_o|$$

$$\phi = \phi_\ell - 2\beta d$$

$$Z(d) = Z_0 \left[\frac{1 + |\Gamma| e^{j\phi}}{1 - |\Gamma| e^{j\phi}} \right]$$

$$= Z_0 \left[\frac{1 + |\Gamma|\cos\phi + j|\Gamma|\sin\phi}{1 - |\Gamma|\cos\phi - j|\Gamma|\sin\phi} \right] \qquad (1)$$

$$= Z_0 \left[\frac{1 - |\Gamma|^2\cos^2\phi - |\Gamma|^2\sin^2\phi + j2|\Gamma|\sin\phi}{(1 - |\Gamma|\cos\phi)^2 + (|\Gamma|\sin\phi)^2} \right]$$

OR

$$Z(d) = Z_0 \left[\frac{1 - |\Gamma|^2 + j2|\Gamma|\sin\phi}{1 - 2|\Gamma|\cos\phi + |\Gamma|^2} \right] = R(d) + jX(d)$$

HENCE: $R(d) = Z_0 \dfrac{1 - |\Gamma|^2}{1 - 2|\Gamma|\cos\phi + |\Gamma|^2}$, $X(d) = Z_0 \dfrac{2|\Gamma|\sin\phi}{1 - 2|\Gamma|\cos\phi + |\Gamma|^2}$

ALSO, FROM (1)

$$|Z(d)| = Z_0 \left[\frac{(1 + |\Gamma|\cos\phi)^2 + (|\Gamma|\sin\phi)^2}{(1 - |\Gamma|\cos\phi)^2 + (|\Gamma|\sin\phi)^2} \right]^{1/2} = Z_0 \left[\frac{1 + 2|\Gamma|\cos\phi + |\Gamma|^2}{1 - 2|\Gamma|\cos\phi + |\Gamma|^2} \right]^{1/2}$$

$$\theta_d = \tan^{-1} \frac{X(d)}{R(d)} = \tan^{-1} \left[\frac{2|\Gamma|\sin\phi}{1 - |\Gamma|^2} \right]$$

2.5)

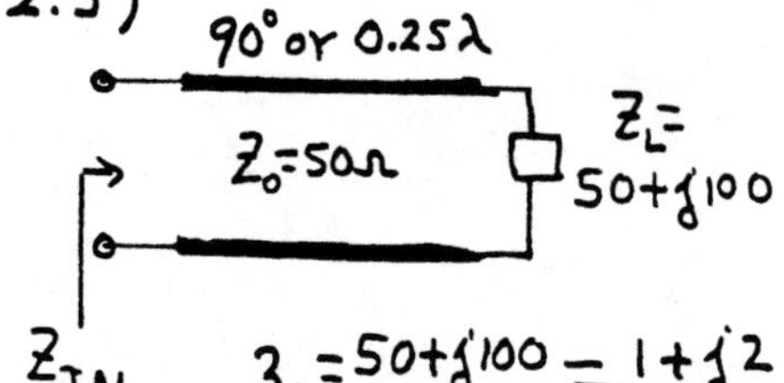

$z_L = \dfrac{50 + j100}{50} = 1 + j2$

LOCATE z_L IN THE SMITH CHART. AT z_L READ 0.187λ. DRAW A CONSTANT $|\Gamma|$ CIRCLE. z_{IN} IS AT 0.25λ FROM z_L (OR 0.187λ + 0.25λ = 0.437λ). HENCE: $z_{IN} = 0.21 - j0.41$

$Z_{IN} = 50 z_{IN} = 10.5 - j20.5\,\Omega$

$\Gamma_0 = 0.707 \angle 45^\circ$

$\text{VSWR} = \dfrac{1 + 0.707}{1 - 0.707} = 5.83$

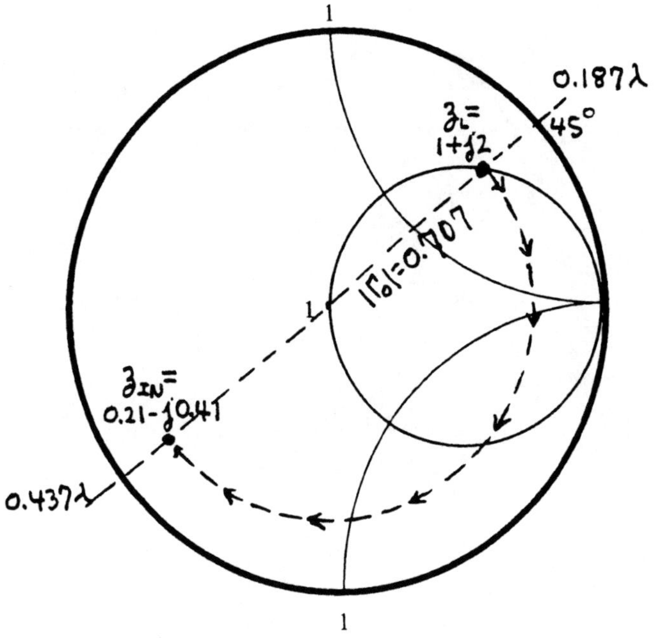

2.6) (a) $|V(d)|_{max} = |A_1|(1+|\Gamma_0|)$ AND $|I(d)|_{min} = \frac{|A_1|}{Z_0}(1-|\Gamma_0|)$

HENCE: $R(d)_{max} = \frac{|V(d)|_{max}}{|I(d)|_{min}} = Z_0 \frac{1+|\Gamma_0|}{1-|\Gamma_0|}$

$$r(d)_{max} = \frac{R(d)_{max}}{Z_0} = \frac{1+|\Gamma_0|}{1-|\Gamma_0|} = VSWR$$

(b) $|V(d)|_{min} = |A_1|(1-|\Gamma_0|)$ AND $|I(d)|_{max} = \frac{|A_1|}{Z_0}(1+|\Gamma_0|)$

HENCE: $R(d)_{min} = \frac{|V(d)|_{min}}{|I(d)|_{max}} = Z_0 \frac{1-|\Gamma_0|}{1+|\Gamma_0|}$

$$r(d)_{min} = \frac{R(d)_{min}}{Z_0} = \frac{1-|\Gamma_0|}{1+|\Gamma_0|} = \frac{1}{VSWR}$$

2.7) (a) $z_{IN} = -j\frac{25}{50} = -j0.5$

FROM THE Z SMITH CHART : $l = 0.426\lambda$

(b) $y_{IN} = j2$

FROM THE Y SMITH CHART: $l = 0.176\lambda$

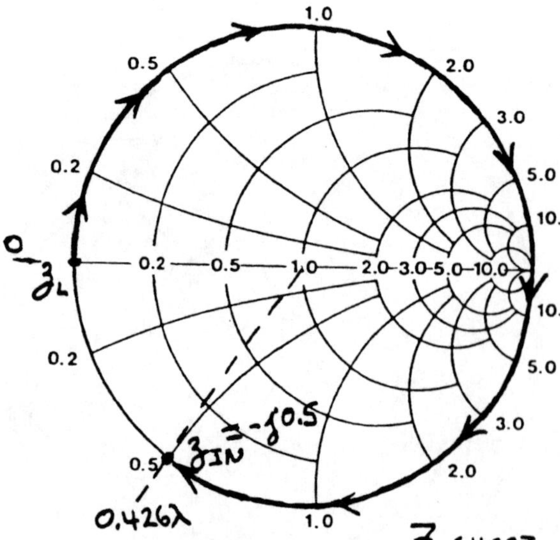

Z CHART

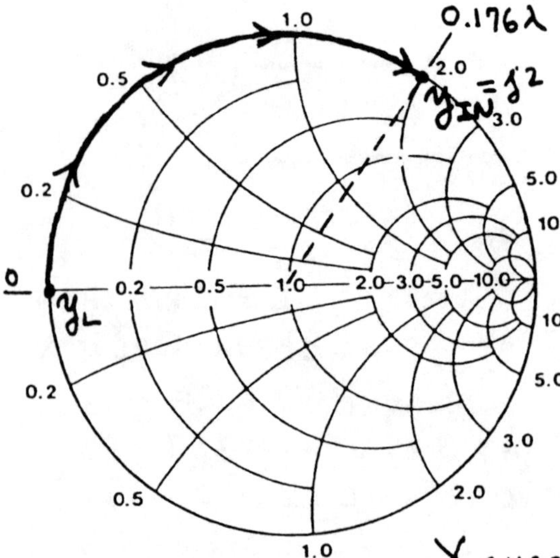

Y CHART

2.8) THE FREQUENCY RESPONSE FOLLOWS A CONSTANT CONDUCTANCE CIRCLE WITH $g=1$. THE EQUIVALENT CIRCUIT IS (SEE FIG. 2.3.2b)

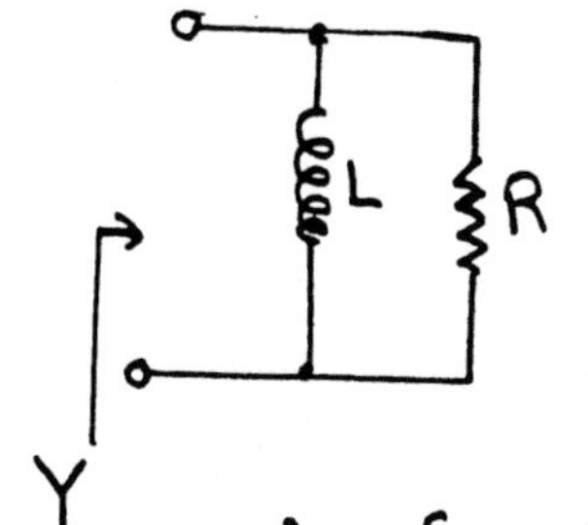

$$Y = \frac{1}{R} - \frac{j}{\omega L}$$

$$y = \frac{Y}{Y_o} = YZ_o = \frac{Z_o}{R} - j\frac{Z_o}{\omega L}$$

$$g = 1 = \frac{Z_o}{R} \Rightarrow R = 50\,\Omega$$

AT $f_b = 1\text{GHz}$: $-j\frac{50}{\omega_b L} = -j0.5$, $L = \frac{50}{0.5(2\pi 10^9)} = 15.9\,\text{nH}$

OBSERVE THAT AT $f_a = 500\text{MHz}$: $y = -j\frac{Z_o}{\omega_a L} = \frac{-j50}{2\pi\, 500\, 10^6\, 15.9\, 10^{-9}} = j1$

2.9) (a) $r = 0.4$, $R = 50r = 50(0.4) = 20\,\Omega$

AT f_a, THE REACTANCE IS $-j0.6$; AND AT f_b IS $-j0.32$.

$$Z = 20 - \frac{j}{\omega C} \qquad \therefore \frac{1}{\omega_a C} = 0.6(50) = 30 \qquad (C = 50\,\text{pF})$$

OR $f_a = \frac{1}{2\pi(30)\, 50\, 10^{-12}} = 106.1\,\text{MHz}$

AND

$$\frac{1}{\omega_b C} = 0.32(50) = 16 \Rightarrow f_b = \frac{1}{2\pi(16)\, 50\, 10^{-12}} = 198.9\,\text{MHz}$$

(b) THE EQUIVALENT CIRCUIT IS A SERIES R, L, C CIRCUIT.

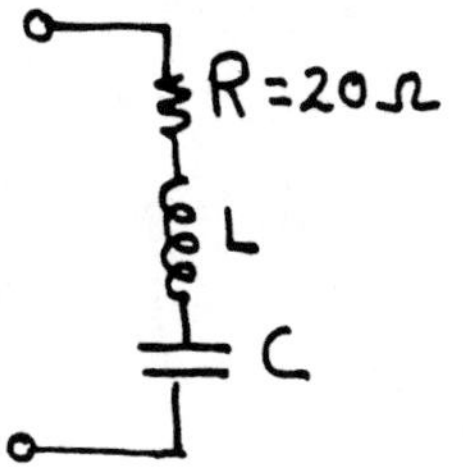

AT $f_a = 500\,\text{MHz}$, $Z = 20 + j10$

$$\therefore 20 + j10 = R + j\left(\omega_a L - \frac{1}{\omega_a C}\right) \qquad (1)$$

AT $f_b = 1\,\text{GHz}$, $Z = 20 + j30$

$$\therefore 20 + j30 = R + j\left(\omega_b L - \frac{1}{\omega_b C}\right) \qquad (2)$$

FROM (1) AND (2) IT FOLLOWS THAT $R = 20\,\Omega$ AND

$$10 = 2\pi\left(f_a L - \frac{1}{f_a C}\right) \quad \text{AND} \quad 30 = 2\pi\left(f_b L - \frac{1}{f_b C}\right)$$

THE SIMULTANEOUS SOLUTION OF THESE EQUATIONS IS:

$L = 5.31\,\text{nH}$

$C = 47.64\,\text{pF}$

2.10)

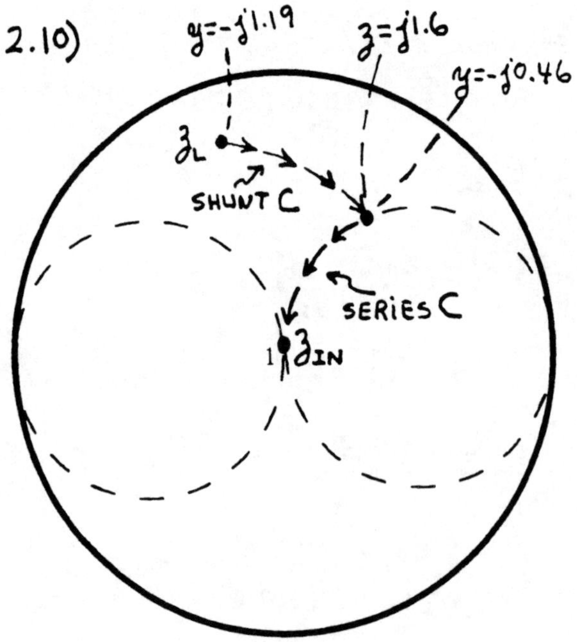

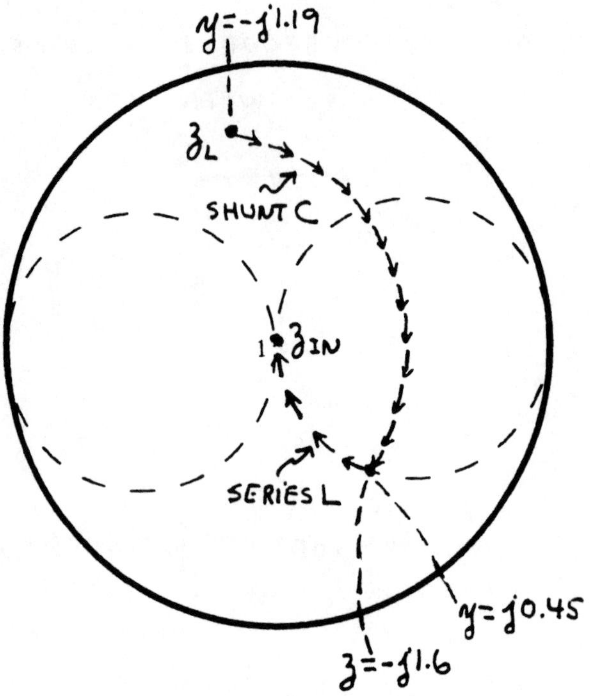

SHUNT C: $y_c = -j0.46 - (-j1.19) = j0.73$

$Z_c = \frac{50}{y_c} = -j68.5\,\Omega$

SERIES C: $z_c = 0 - j1.6 = -j1.6$

$Z_c = 50 z_c = -j80\,\Omega$

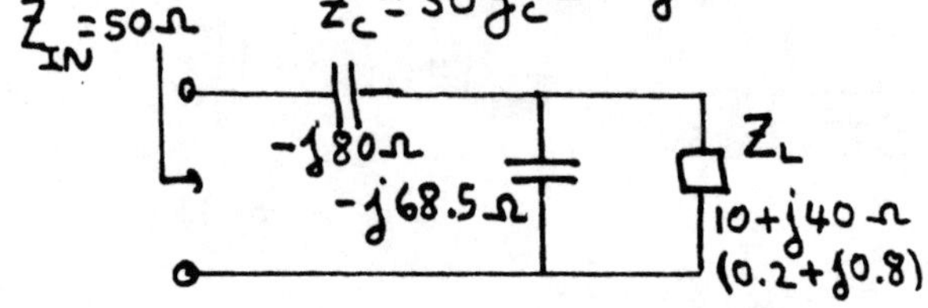

SHUNT C: $y_c = j0.45 - (-j1.19) = j1.64$

$Z_c = \frac{50}{j1.64} = -j30.5\,\Omega$

SERIES L: $z_L = 0 - (-j1.6) = j1.6$

$Z_L = 50 z_L = j80\,\Omega$

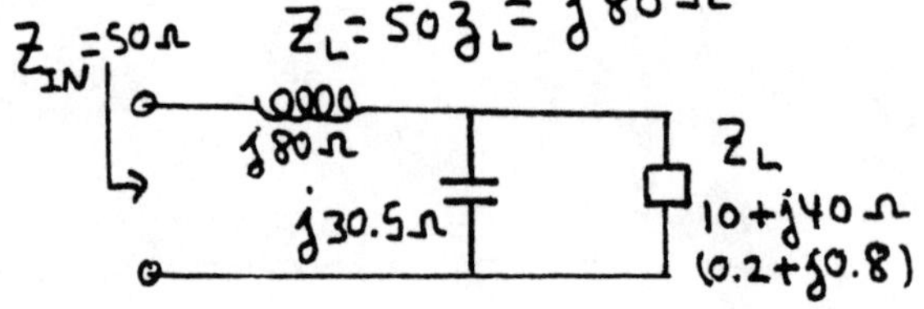

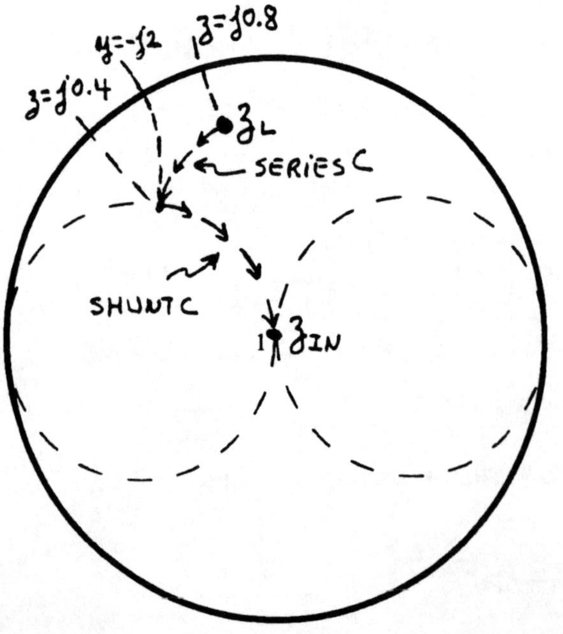

SERIES C: $z_c = j0.4 - j0.8 = -j0.4$

$Z_c = 50 z_c = -j20\,\Omega$

SHUNT C: $y_c = 0 - (-j2) = j2$

$Z_c = \frac{50}{j2} = -j25\,\Omega$

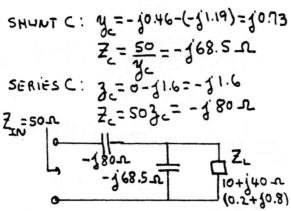

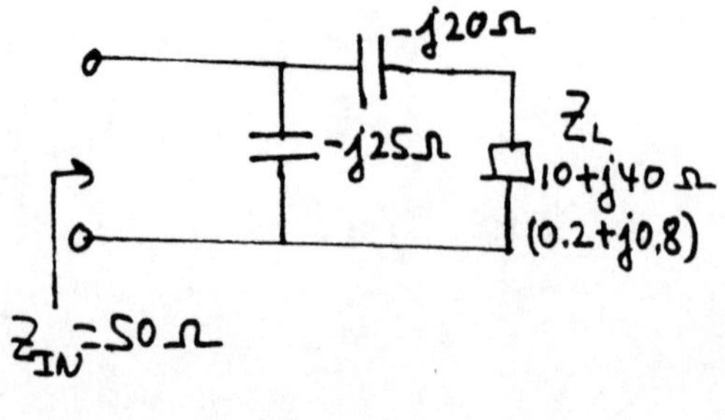

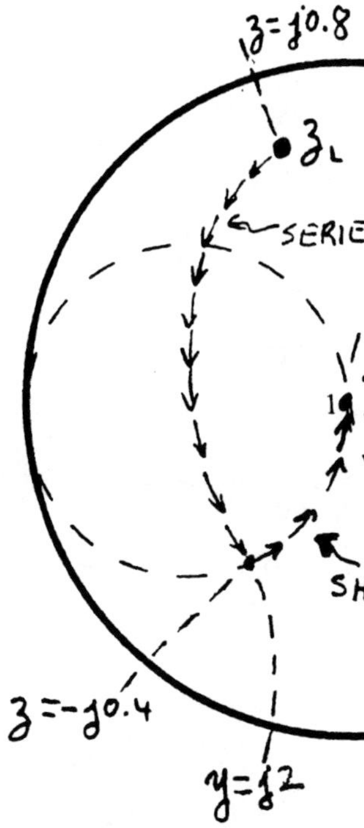

SERIES C: $z_C = -j0.4 - j0.8 = -j1.2$

$Z_C = 50 z_C = -j60\ \Omega$

SHUNT L: $y_L = 0 - j2 = -j2$

$Z_L = \frac{50}{y_L} = j25\ \Omega$

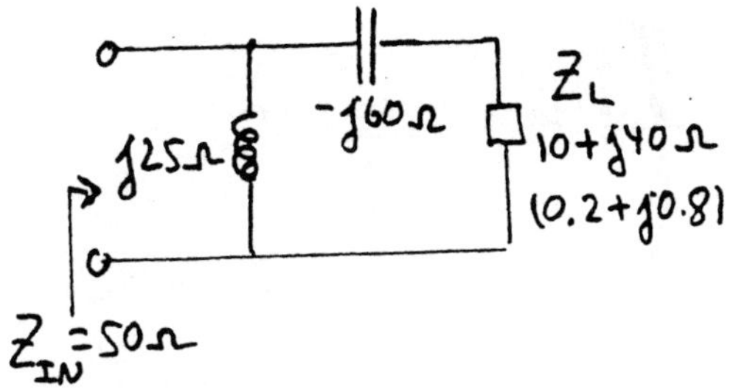

2.11) $y_L = \frac{Y_L}{Y_0} = Z_0 Y_L = 50(4-j4)10^{-3} = 0.2 - j0.2$

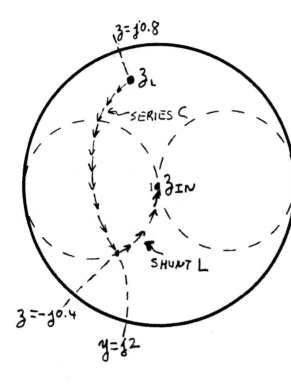

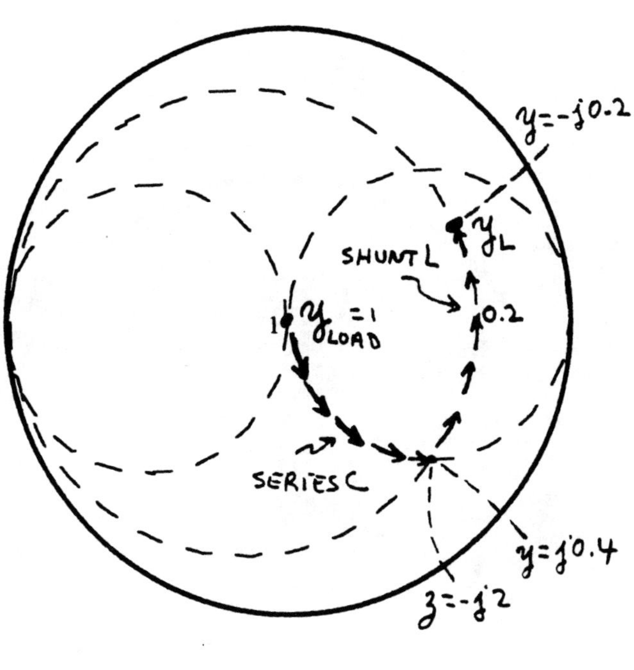

SERIES C: $z_C = -j2 - j0 = -j2$

$Z_C = 50 z_C = -j100\Omega$

SHUNT L: $y_L = -j0.2 - j0.4 = -j0.6$

$Z_L = \frac{50}{y_L} = j83.3\Omega$

AT $f = 700$ MHz:

$Z_L = j\omega L = j83.3$

$L = \frac{83.3}{2\pi\, 700 10^6} = 18.9$ nH

$Z_C = \frac{-j}{\omega C} = -j100$

$C = \frac{1}{100(2\pi\, 700 10^6)} = 2.27$ pF

2.12) ONLY CIRCUIT (b) CAN MATCH $Y_L = 8 - j12$ mS to $Z_{IN} = 50\,\Omega$.

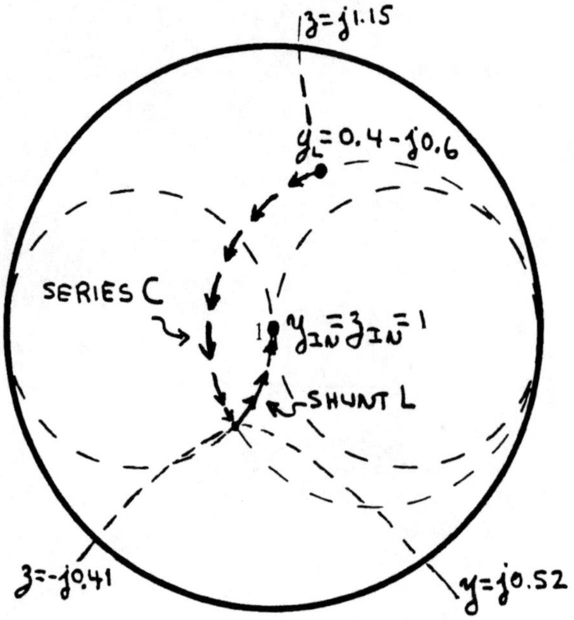

SERIES C: $z_C = -j0.41 - j1.15 = -j1.56$

$Z_C = 50 z_C = -j78\,\Omega$

$C = \frac{1}{\omega(78)} = \frac{1}{2\pi 10^9 (78)} = 2.04$ pF

SHUNT L: $y_L = 0 - j0.52 = -j0.52$

$Z_L = \frac{50}{y_L} = j96.15\,\Omega$

$L = \frac{96.15}{\omega} = \frac{96.15}{2\pi 10^9} = 15.3$ nH

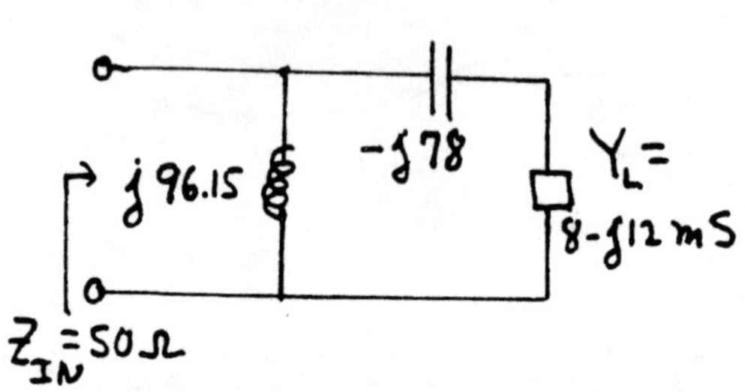

2.13) $Z_L = j\omega L = j10^9\, 50\, 10^{-9} = j50$ OR $z_L = j\frac{50}{50} = j1$

$Z_C = \frac{1}{j\omega C} = \frac{1}{j10^9\, 10\, 10^{-12}} = -j100$ OR $z_C = -j\frac{100}{50} = -j2$

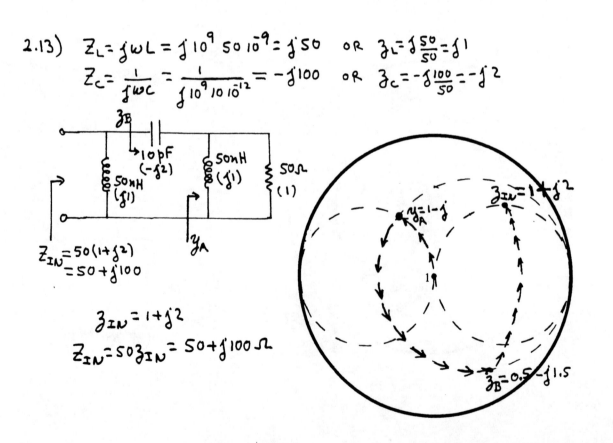

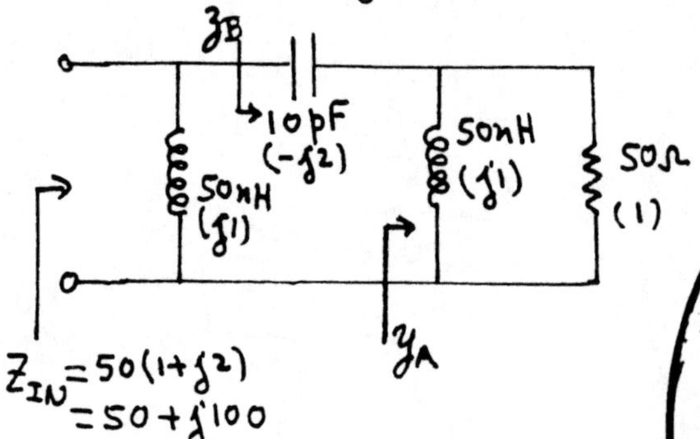

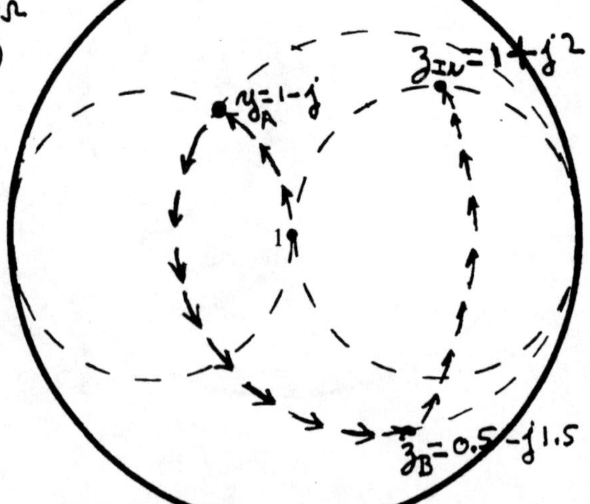

$z_{IN} = 1 + j2$

$Z_{IN} = 50 z_{IN} = 50 + j100\,\Omega$

2.14) LET $Z_0 = 100\Omega$.

THEN $z_L = \frac{Z_L}{100} = 1 - j$

AND $z_{IN} = \frac{Z_{IN}}{100} = 0.25 + j0.25$

SHUNT C: $y_C = j1.31 - j0.5 = j1.26$

$$Z_C = \frac{100}{y_C} = -j126\,\Omega$$

SERIES L: $z_L = j0.25 - (-j0.67) = j0.92$

$$Z_L = 100 z_L = j92\,\Omega$$

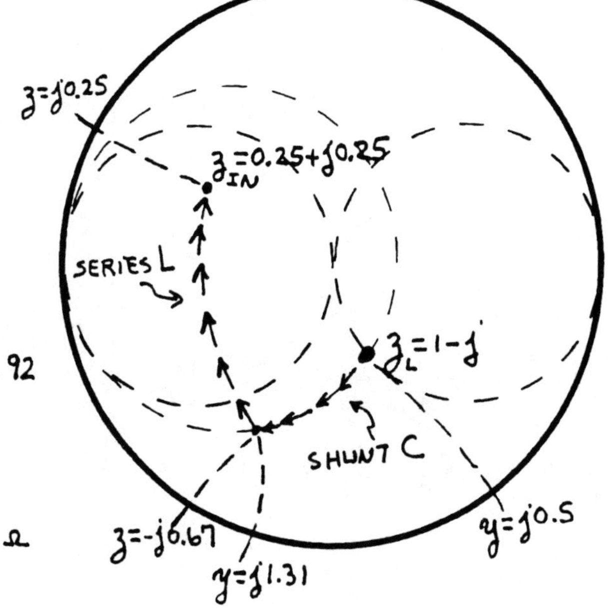

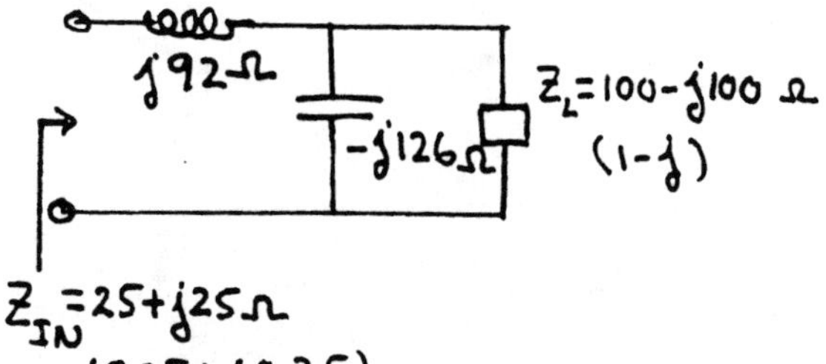

2.15) (a) $z_L = \frac{50}{50} = 1$

$z_{IN} = \frac{20 + j20}{50} = 0.4 + j0.4$

DRAW THE Q=5 CIRCLES (SEE FIG. 2.4.16)

THE MOTION FROM A TO B -- SERIES L_1:

AT B: $z_B = 1 + j3$

$z_{L_1} = j3$ or $Z_{L_1} = j3(50) = j150\,\Omega$

THE MOTION FROM B TO C -- SHUNT C:

AT B: $y_B = 0.1 - j0.3$

AT C: $y_C = 0.1 + j0.5$

$y_C = j0.5 - (-j0.3) = j0.8$

$$Z_C = \frac{50}{j0.8} = -j62.5\,\Omega$$

THE MOTION FROM C TO D -- SERIES L_2:

AT C: $z_C = 0.4 - j1.9$

AT D: $z_D = 0.4 + j0.4$

$z_{L_2} = j0.4 - (-j1.9) = j2.3$

$$Z_{L_2} = 50(j2.3) = j115\,\Omega$$

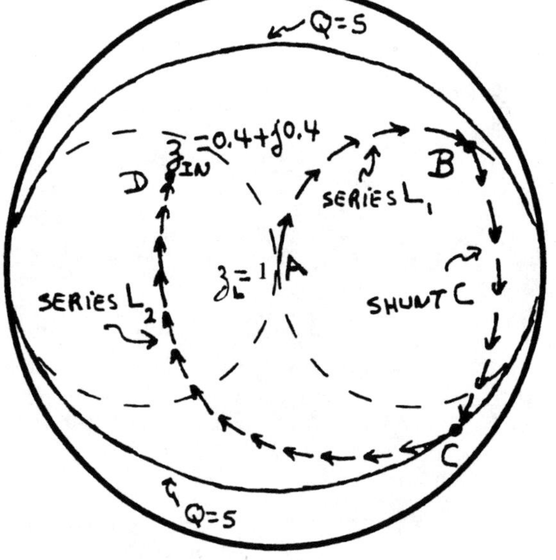

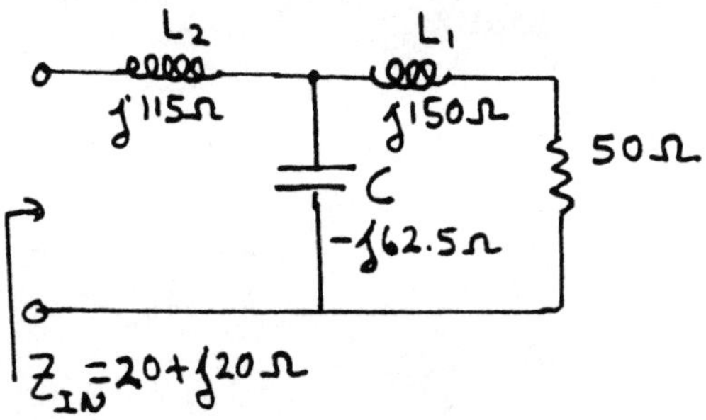

(b) $z_L = 1$ AND $z_{IN} = 0.5$

AT B: $y_B = 1 - j2.6$, $z_B = 0.13 + j0.335$

AT C: $y_C = 2 - j3.4$, $z_C = 0.13 + j0.215$

AT D: $y_D = 2$, $z_D = z_{IN} = 0.5$

SHUNT L: $y_L = -j2.6$

$$Z_L = \frac{50}{-j2.6} = j19.2\ \Omega$$

SERIES C_1: $z_{C_1} = j0.215 - j0.335 = -j0.12$

$$Z_{C_1} = 50(-j0.12) = -j6\ \Omega$$

SHUNT C_2: $y_{C_2} = 0 - (-j3.4) = j3.4$

$$Z_{C_2} = \frac{50}{j3.4} = -j14.7\ \Omega$$

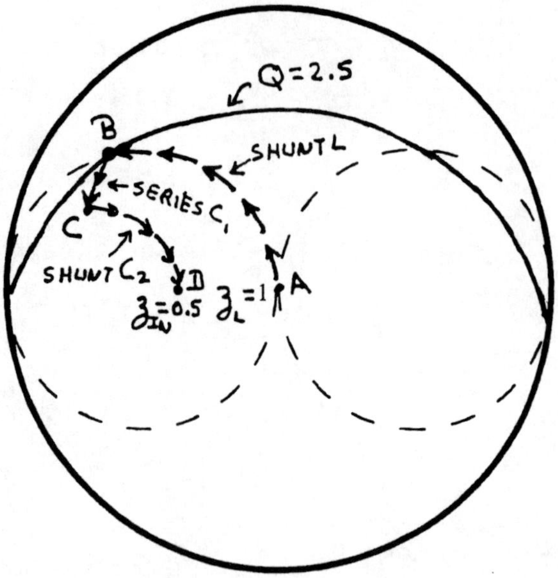

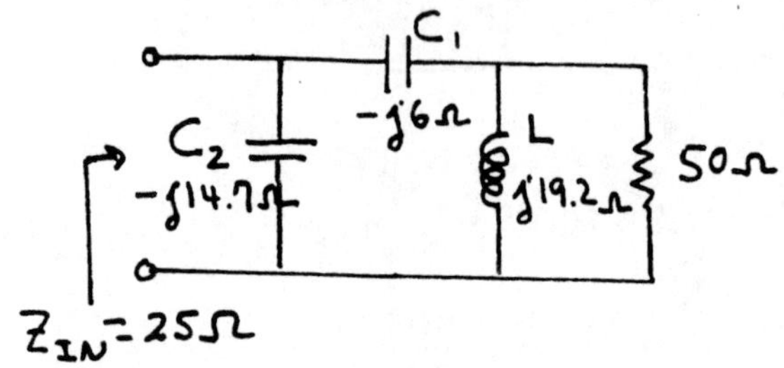

2.16) (a) FROM FIG. 2.5.2 WITH $Z_0 = 50\ \Omega$ AND $\epsilon_r = 2.23$ WE OBTAIN:

$$\frac{W}{h} \simeq 3.1 \quad \text{OR} \quad W = 3.1(0.7874) = 2.44\ \text{mm}$$

FROM FIG. 2.5.3 WITH $W/h = 3.1$ AND $\epsilon_r = 2.23$ WE OBTAIN:

$$\frac{\lambda}{\lambda_{TEM}} = 1.08 \quad \text{OR} \quad \lambda = 1.08\lambda_{TEM} = 1.08\frac{\lambda_0}{\sqrt{2.23}} = 0.723\lambda_0$$

SINCE $\lambda = \dfrac{\lambda_0}{\sqrt{\epsilon_{ff}}}$, THEN $\dfrac{1}{\sqrt{\epsilon_{ff}}} = 0.723$ OR $\epsilon_{ff} = 1.91$

(b) FROM (2.5.11): $\dfrac{W}{h} = \dfrac{2}{\pi}\left\{B - 1 - \ln(2B-1) + \dfrac{2.23-1}{2(2.23)}\left[\ln(B-1) + 0.39 - \dfrac{0.61}{2.23}\right]\right\}$ (1)

WHERE $B = \dfrac{377\pi}{2(50)\sqrt{2.23}} = 7.931$ (2)

SUBSTITUTE (2) INTO (1) TO OBTAIN: $\dfrac{W}{h} = 3.073$

FROM (2.5.8): $\lambda = \dfrac{\lambda_0}{\sqrt{2.23}}\left[\dfrac{2.23}{1 + 0.63(2.23-1)(3.073)^{0.1255}}\right]^{1/2} = 0.724\lambda_0$

$$\frac{1}{\sqrt{\epsilon_{ff}}} = 0.724 \quad \text{OR} \quad \epsilon_{ff} = 1.91$$

2.17) APPROXIMATE VALUES CAN BE OBTAINED FROM FIGS. 2.5.2 AND 2.5.3. EXACT VALUES ARE GIVEN IN FIG. 2.5.4.

$$\therefore \frac{W}{h} = 1.5 \quad \text{OR} \quad W = 1.5(25) = 37.5 \text{ mils}$$

FROM FIG. 2.5.3: $\frac{\lambda}{\lambda_{TEM}} = 1.18$ OR $\lambda = 1.18\,\lambda_{TEM} = \frac{1.18}{\sqrt{6}}\frac{3\,10^{10}}{10^9} = 14.45 \text{ cm}$

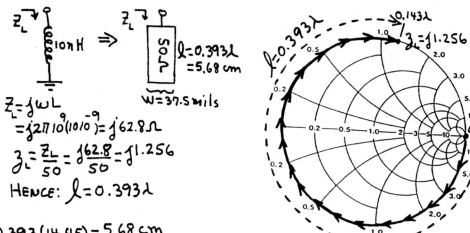

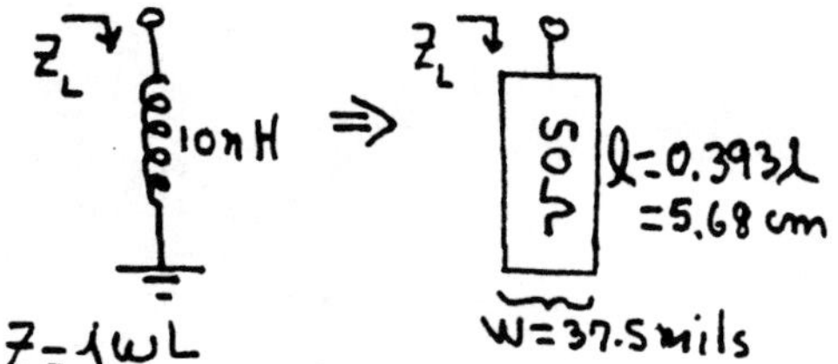

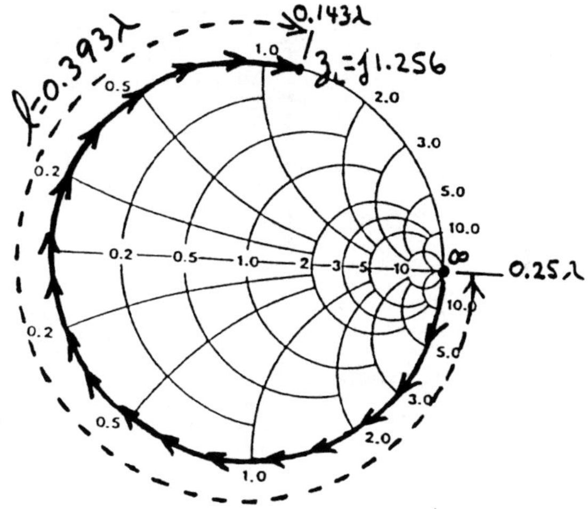

$$Z_L = j\omega L = j2\pi 10^9 (10\,10^{-9}) = j62.8\,\Omega$$

$$z_L = \frac{Z_L}{50} = \frac{j62.8}{50} = j1.256$$

HENCE: $\ell = 0.393\lambda$

OR

$\ell = 0.393(14.45) = 5.68$ cm

OR $\ell = 5.68\left(\frac{1000}{2.54}\right) = 2{,}236.2$ mils

2.18) AT $f = 1$ GHz, $\lambda_o = 30$ cm. MICROSTRIP: $\varepsilon_r = 2.23$, $h = 0.7874$ mm

LINE WITH $Z_0 = 29.9\,\Omega$ AND $\ell = \frac{\lambda}{4}$: $B = \frac{377\pi}{2(29.9)\sqrt{2.23}} = 13.26$

$$\frac{W}{h} = \frac{2}{\pi}\left\{13.26 - 1 - \ln(2(13.26)-1) + \frac{2.23-1}{2(2.23)}\left[\ln(13.26-1) + 0.39 - \frac{0.61}{2.23}\right]\right\} = 6.203$$

$$\therefore W = 6.203h = 6.203(0.7874) = 4.884 \text{ mm}$$

$$\lambda = \frac{\lambda_o}{\sqrt{2.23}}\left[\frac{2.23}{1 + 0.63(2.23-1)(6.203)^{0.1255}}\right]^{1/2} = 0.712\lambda_o$$

NOTE: SINCE $\lambda = \frac{\lambda_o}{\sqrt{\varepsilon_{ff}}}$ THEN $\frac{1}{\sqrt{\varepsilon_{ff}}} = 0.712$ OR $\varepsilon_{ff} = 1.974$

$$\ell = \frac{\lambda}{4} = 0.25(0.712\lambda_o) = 0.25(0.712)(30) = 5.34 \text{ cm}$$

SIMILARLY, WE OBTAINED:

$Z_0 = 52.64\,\Omega$, $\ell = 3\lambda/8$	$Z_0 = 95.2\,\Omega$, $\ell = \frac{3\lambda}{8}$	$Z_0 = 79\,\Omega$, $\ell = \lambda/4$
$\frac{W}{h} = 2.89$ OR $W = 2.27$ mm	$\frac{W}{h} = 0.99$, $W = 0.779$ mm	$\frac{W}{h} = 1.44$, $W = 1.134$ mm
$\lambda = 0.726\lambda_o$, $\varepsilon_{ff} = 1.89$	$\lambda = 0.747\lambda_o$, $\varepsilon_{ff} = 1.79$	$\lambda = 0.74\lambda_o$, $\varepsilon_{ff} = 1.825$
$\ell = \frac{3\lambda}{8} = \frac{3}{8}(0.726)(30) = 8.167$ cm	$\ell = \frac{3\lambda}{8} = 8.4$ cm	$\ell = \frac{\lambda}{4} = 5.55$ cm

2.19) (a) $z_{IN} = \frac{Z_{IN}}{50} = 2 - j2$

$y_{IN} = \frac{1}{z_{IN}} = 0.25 + j0.25$

$l_1 = 0.339\lambda - 0.25\lambda = 0.089\lambda$

$l_2 = 0.042\lambda + (0.5\lambda - 0.321\lambda) = 0.219\lambda$

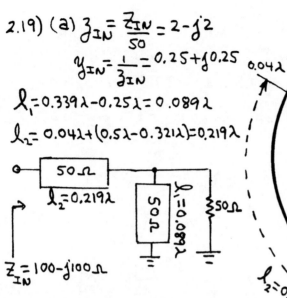

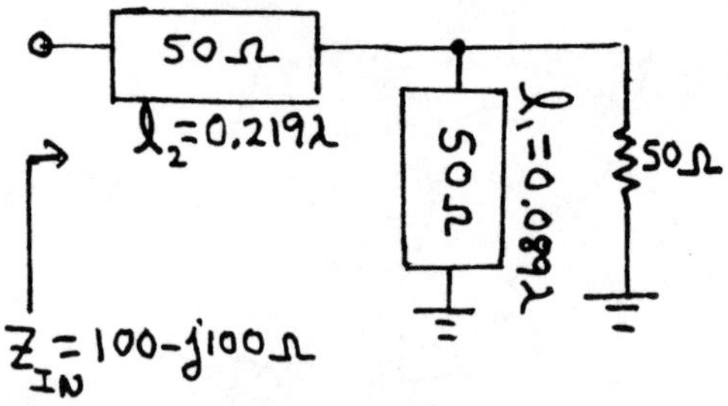

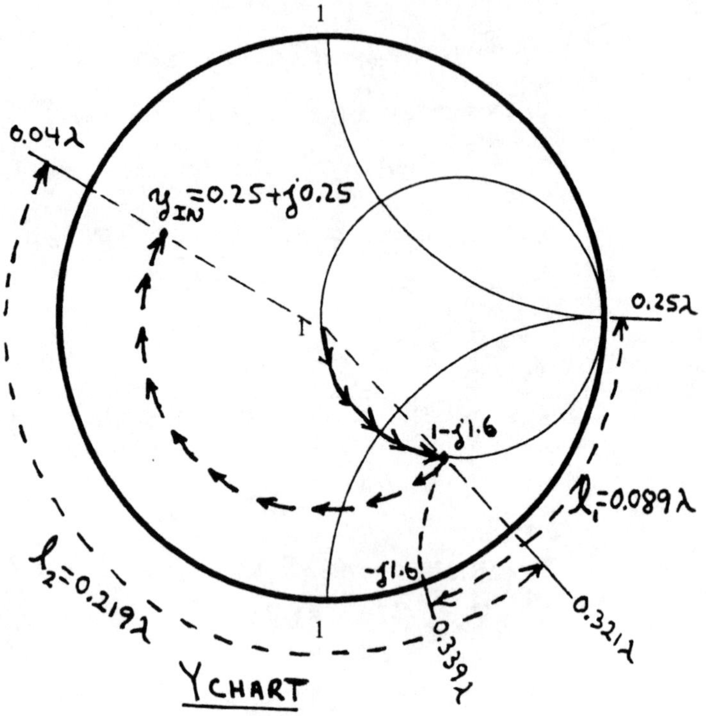

(b) If l_1 is an open-circuited stub, then

$l_1 = 0.25\lambda + 0.089\lambda = 0.339\lambda$

(c) For $z_{IN} = \frac{Z_{IN}}{50} = 2 + j2$, one answer is:

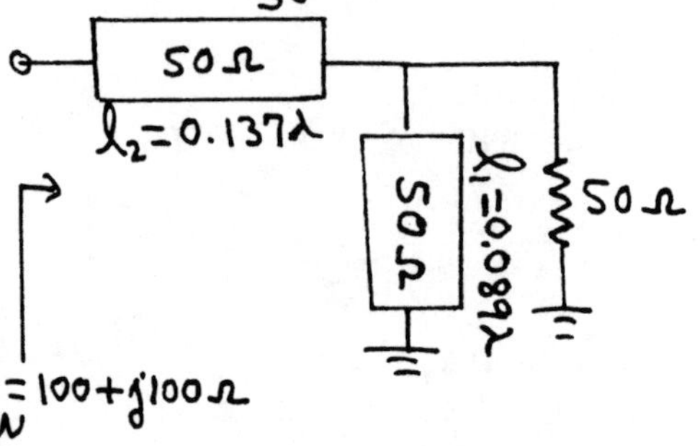

If l_1 is an open-circuited stub, then

$l_1 = 0.25\lambda + 0.089\lambda = 0.339\lambda$

2.20) (a) $z_L = \frac{15 + j25}{50} = 0.3 + j0.5$

$y_L = \frac{1}{z_L} = 0.882 - j1.47$

Rotate, along a constant $|\Gamma|$ circle, from y_L until the unit conductance circle is reached at y_x.

$y_x = 1 + j1.55$

$l_1 = 0.178\lambda + (0.5\lambda - 0.329\lambda)$

$= 0.349\lambda$

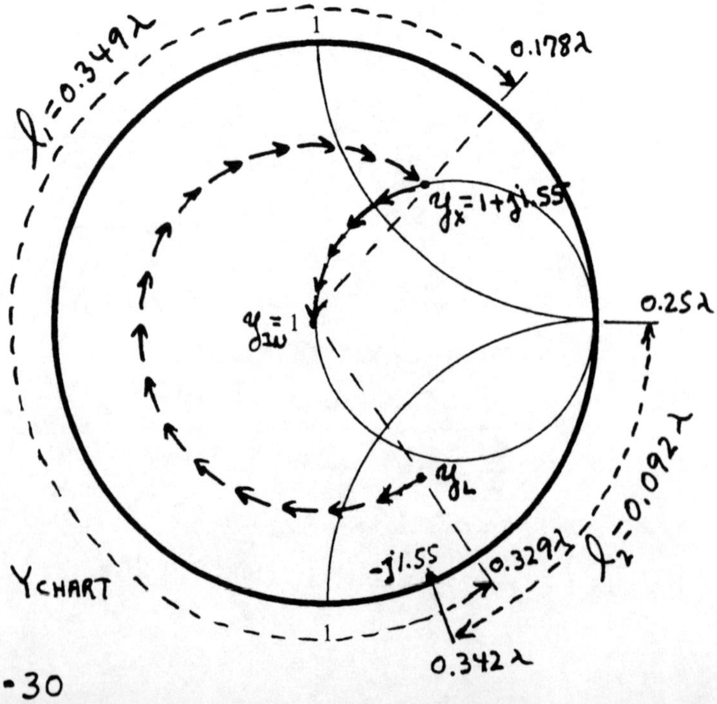

$y_{IN} = y_{s.c} + y_x$

$1 = y_{s.c} + (1 + j1.55)$

HENCE, $y_{s.c} = -j1.55$

The LENGTH l_2 OF THE SHORT CIRCUITED STUB MUST HAVE

$y_{sc} = -j1.55$

FROM THE Y SMITH CHART:

$l_2 = 0.342\lambda - 0.25\lambda = 0.092\lambda$

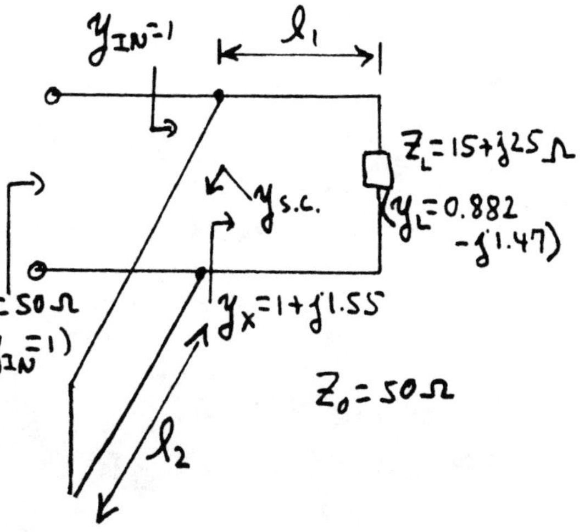

(b) IF THE CHARACTERISTIC IMPEDANCE OF THE STUB IS $Z_0' = 100\,\Omega$.

$$Z_{s.c.} = \frac{50}{y_{s.c}} = \frac{50}{-j1.55} = j32.26\,\Omega$$

THEN: $j32.26 = jZ_0' \tan\beta d = j100 \tan\beta d \Rightarrow \beta d = 0.312$

$$d \equiv l_2 = \frac{0.312}{2\pi/\lambda} = 0.05\lambda$$

ANOTHER WAY: NORMALIZE $Z_{s.c}$ WITH Z_0', $z_{s.c} = j\frac{32.26}{100} = j0.323$

THEN: $y_{s.c} = \frac{1}{z_{s.c}} = -j3.09$. LOCATE $y_{s.c} = -j3.09$ IN THE Y SMITH CHART, AND READ $l_2 = 0.302\lambda - 0.25\lambda = 0.052\lambda$

2.21) (a) $Y_{IN} = G_{IN} + jB_{IN} = 50 + j40\text{ mS}$, $R_{IN} = \frac{1}{G_{IN}} = \frac{1}{50\,10^{-3}} = 20\,\Omega$

$$Z_{01} = \sqrt{Z_L R_{IN}} = \sqrt{50(20)} = 31.62\,\Omega$$

IN A SHORT-CIRCUITED $\frac{3\lambda}{8}$ STUB: $Y_{SC} = jY_{02}$. HENCE, $jY_{02} = jB_{IN} = j40\text{ mS}$

OR $Y_{02} = 40\text{ mS}$, $Z_{02} = \frac{1}{Y_{02}} = 25\,\Omega$

(b) $Y_{IN} = G_{IN} - jB_{IN} = 50 - j40\text{mS}$, $R_{IN} = \frac{1}{G_{IN}} = 20\,\Omega$.

THEN, $Z_{01} = \sqrt{50(20)} = 31.62\,\Omega$. IN A SHORT-CIRCUITED $\frac{\lambda}{8}$ STUB:

$Y_{SC} = -jY_{02}$. HENCE, $Y_{02} = 40\text{mS}$ OR $Z_{02} = \frac{1}{40\,10^{-3}} = 25\,\Omega$.

(c) $Y_{IN} = G_{IN} + jB_{IN} = 10 + j20\text{ mS}$, $R_{IN} = \frac{1}{G_{IN}} = \frac{1}{10\,10^{-3}} = 100\,\Omega$

THEN, $Z_{01} = \sqrt{50(100)} = 70.7\,\Omega$. IN AN OPEN-CIRCUITED $\frac{\lambda}{8}$ STUB:

$Y_{o.c} = jY_{02}$. HENCE, $Y_{02} = 20\text{mS}$ OR $Z_{02} = \frac{1}{20\,10^{-3}} = 50\,\Omega$.

(d) $Y_{IN} = 10 - j20\text{m S}$. HENCE: $Z_{01} = \sqrt{50(100)} = 70.7\,\Omega$.

IN AN OPEN-CIRCUITED $\frac{3\lambda}{8}$ STUB: $Y_{OC} = -jY_{02}$. HENCE, $Z_{02} = \frac{1}{Y_{02}} = \frac{1}{20\,10^{-3}} = 50\,\Omega$.

2.22) (a) $\Gamma_L = 0.5\angle 90^\circ$, $z_L = 0.6 + j0.8$

$\therefore y_L = \frac{1}{z_L} = 0.6 - j0.8$

FROM THE SMITH CHART:

$l_1 = 0.136\lambda$, $l_2 = 0.375\lambda - 0.166\lambda$
$= 0.209\lambda$

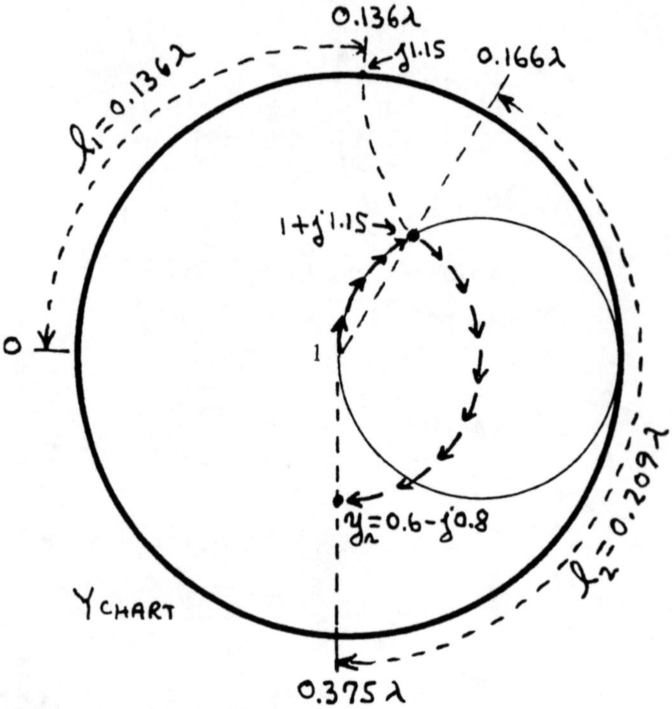

IN FIG. P.22(b):

$Y_L = \frac{0.6 - j0.8}{50} = 12 - j16 \text{ mS}$

$\therefore Z_{01} = \sqrt{50\left(\frac{1}{12 \cdot 10^{-3}}\right)} = 64.5\,\Omega$

USING A $\frac{3\lambda}{8}$ OPEN-CIRCUITED STUB:

$-jY_{02} = -j16\text{mS}$, OR $Y_{02} = 16\text{ mS}$

$Z_{02} = \frac{1}{16 \cdot 10^{-3}} = 62.5\,\Omega$

(b) BALANCED FORM OF THE STUBS.

FOR FIG. P.22(a): $y_{bal} = j\frac{1.15}{2} = j0.575 \Rightarrow l_{1(bal)} = 0.083\lambda$

FOR FIG. P.22(b): $Z_{02(bal)} = 2(62.5) = 125\,\Omega$

2.23) THE $\frac{3\lambda}{8}$ STUB WITH $Z_0 = 26.32\,\Omega$ HAS AN ADMITTANCE OF $-j0.038S$.

IN A $Z_0 = 26.32\,\Omega$ SYSTEM: $y_{oc} = (26.32)(-j0.038) = -j1$

THEN: $y_{oc(bal)} = \frac{-j1}{2} = -j0.5 \Rightarrow l_{(bal)} = 0.426\lambda$

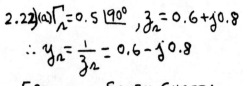

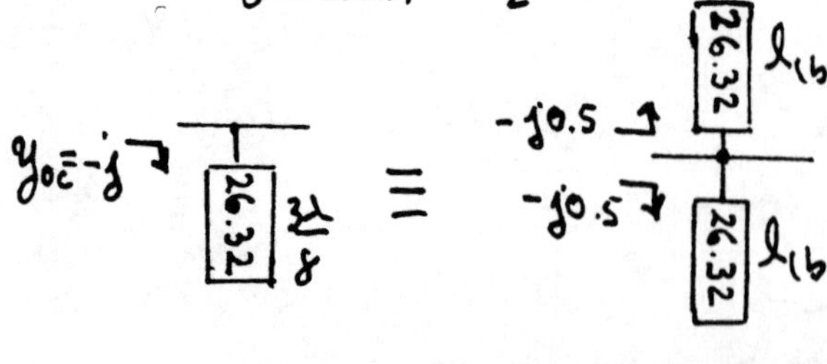

THE $\frac{3\lambda}{8}$ STUB WITH $Z_0 = 47.6\,\Omega$ HAS AN ADMITTANCE OF $-j0.021$ S.

$y_{oc} = 47.6(-j0.021) = -j1$

THEN: $y_{oc(bal)} = -j\frac{1}{2} = -j0.5 \Rightarrow l_{(bal)} = 0.426\lambda$

2.24) (a) $\Gamma_L = 0.4\angle{-120^\circ}$, $z_L = 0.538 - j0.444$, $y_L = \frac{1}{z_L} = 1.105 + j0.912$

$Y_L = \frac{y_L}{50} = 22 + j18 \text{ mS}$

HENCE: $Z_{01} = \sqrt{50\left(\frac{1}{22\,10^{-3}}\right)} = 47.67\,\Omega$ AND

$jY_{02} = j18 \text{ mS}$ OR $Z_{02} = \frac{1}{Y_{02}} = \frac{1}{18\,10^{-3}} = 55.56\,\Omega$

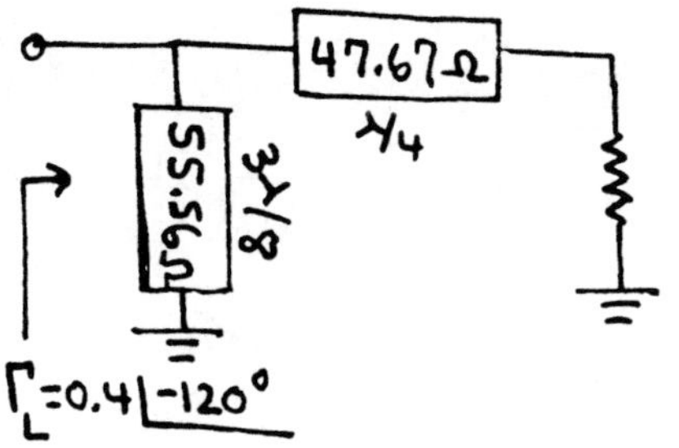

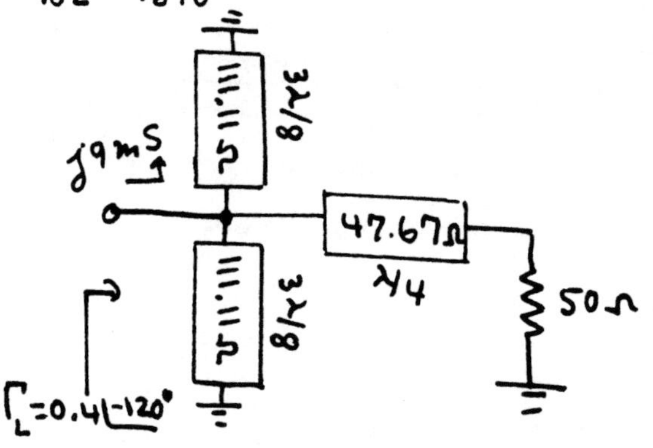

BALANCE FORM (USE $2Z_{02} = 111.11\,\Omega$)

(b) EACH SIDE OF THE BALANCE STUBS HAS AN ADMITTANCE OF $j9\text{mS}$. IF ITS CHARACTERISTIC IMPEDANCE IS $Z_0 = \frac{111.11}{2} = 55.56\,\Omega$, THEN $y_{(b2l)} = j9\,10^{-3}(55.56) = j0.5$. HENCE: $l = 0.323\lambda$.

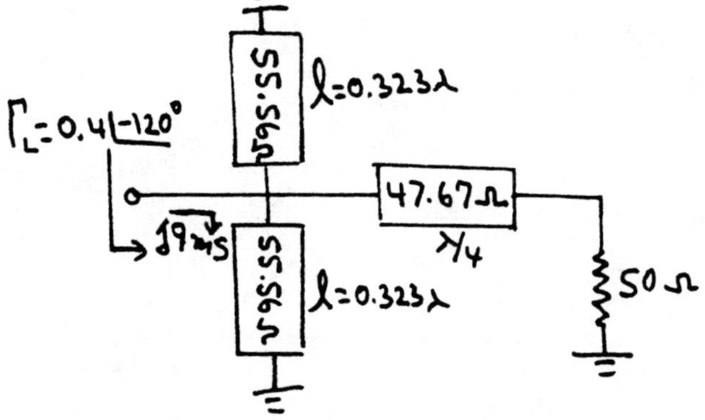

2.25) $\Gamma_s = 0.8\angle{160^\circ}$, $y_s = \frac{1}{z_s} = 2.64 - j4$

$\Gamma_L = 0.7\angle{20^\circ}$, $y_L = \frac{1}{z_L} = 0.182 - j0.171$

THE DESIGN OF THE MATCHING CIRCUITS IS SHOWN IN THE Y SMITH CHARTS.

MATCHING TO Γ_s:

OPEN STUB LENGTH: $l_1 = 0.192\lambda$

SERIES TRANS. LINE: $l_2 = 0.08\lambda$

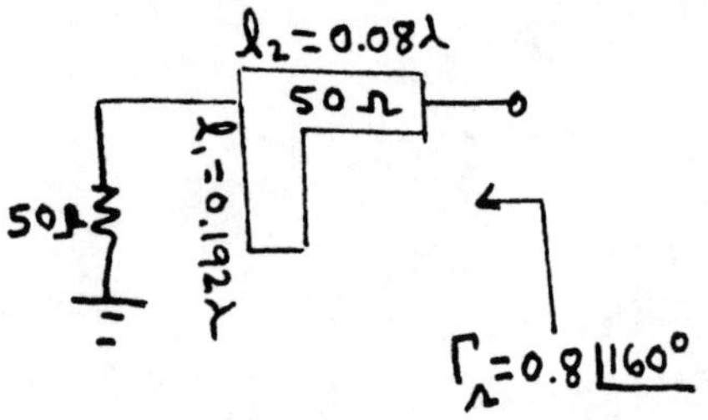

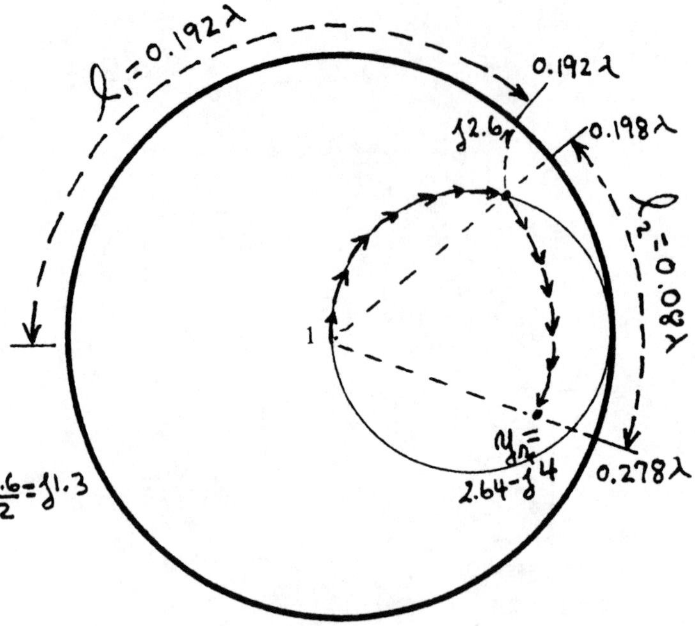

BALANCED SHUNT STUBS: $y_{(bal)} = j\frac{2.6}{2} = j1.3$

LENGTH OF EACH SIDE: $l_1' = 0.146\lambda$

MATCHING TO Γ_L:

SHORT-CIRCUITED STUB LENGTH: $l_1 = 0.076\lambda$

SERIES TRANS. LINE: $l_2 = 0.159\lambda$

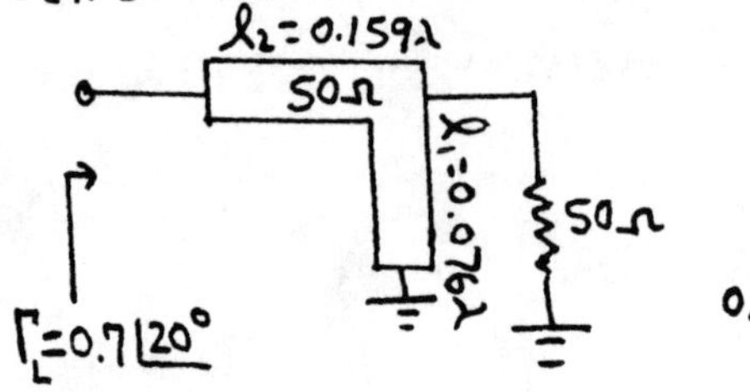

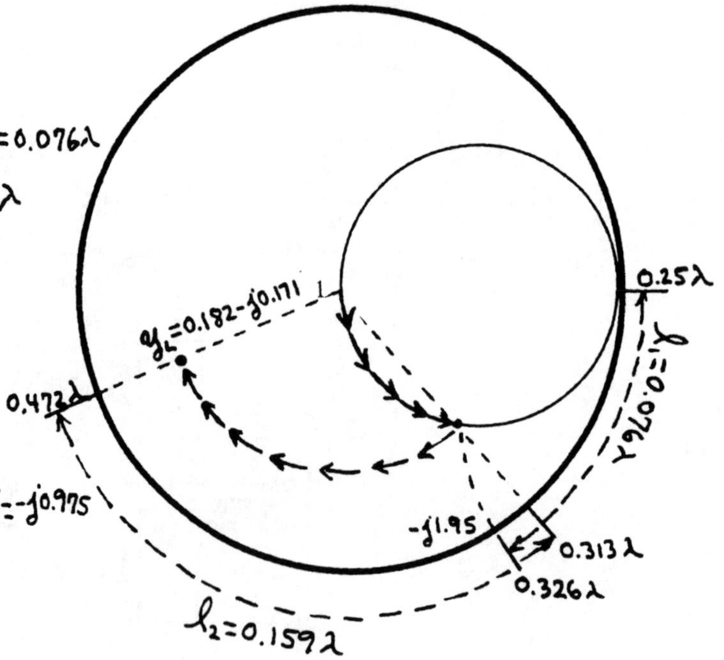

BALANCED SHUNT STUBS: $y_{(bal)} = -j\frac{1.95}{2} = -j0.975$

LENGTH OF EACH SIDE: $l_1' = 0.127\lambda$

AMPLIFIER DIAGRAM

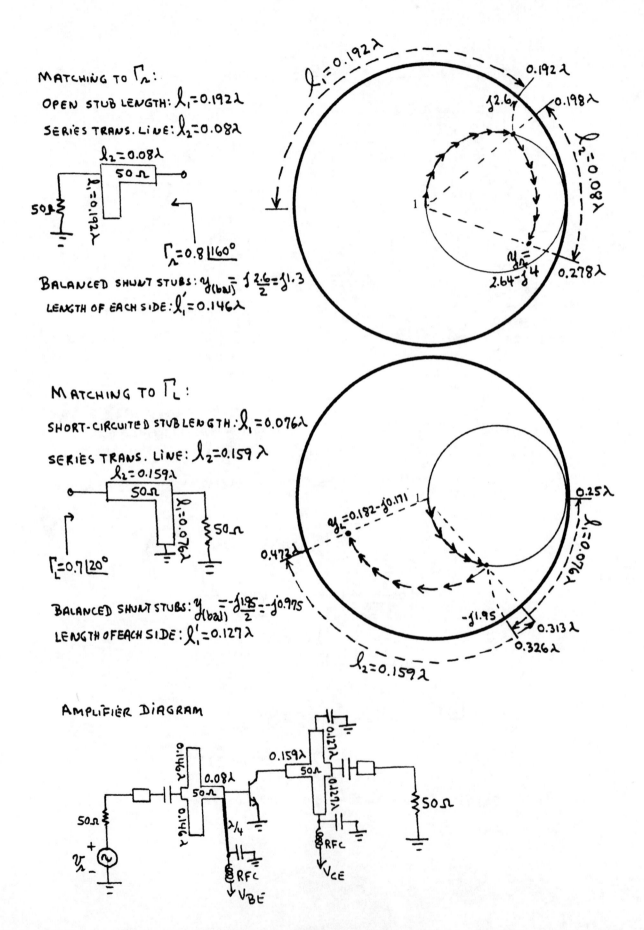

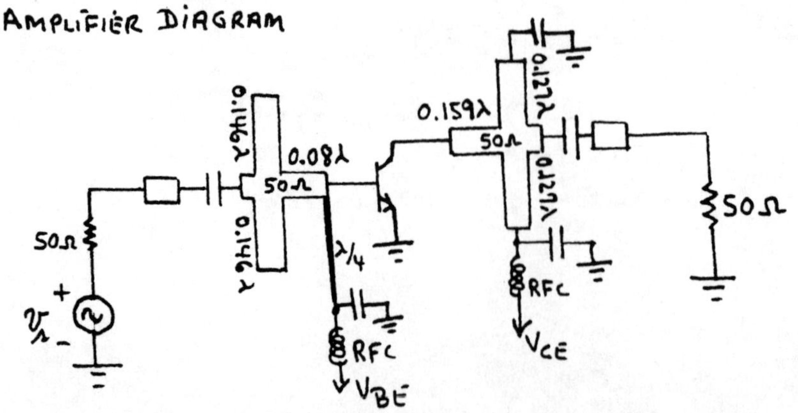

2.26) (a)

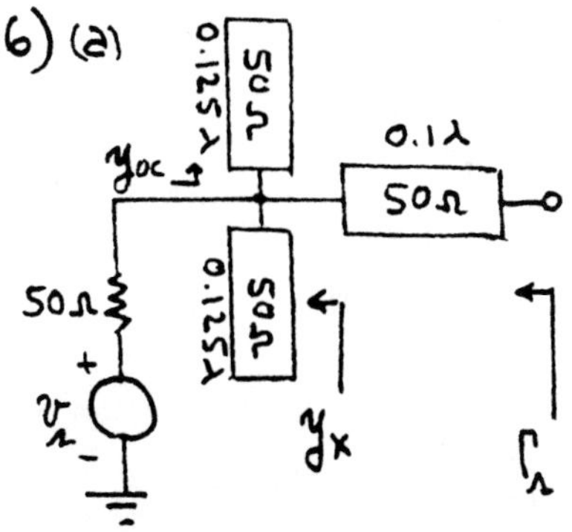

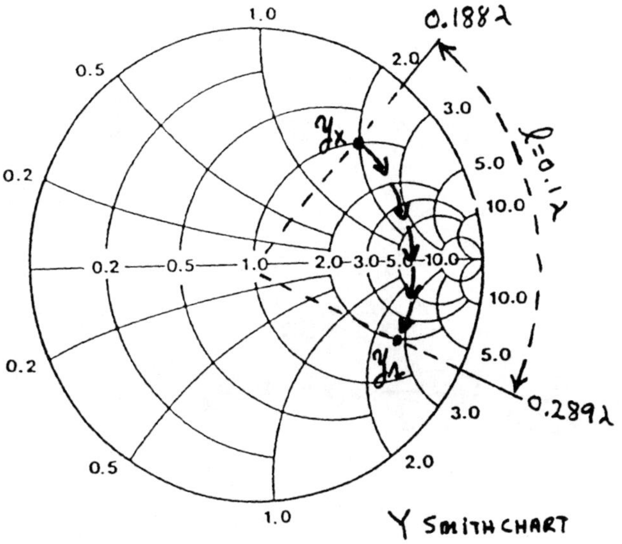

THE ADMITTANCE OF THE 0.125λ STUB IS: $y_{oc} = j$

HENCE: $y_x = 1 + j + j = 1 + 2j$

LOCATE y_x IN THE Y CHART, AND ROTATE 0.1λ TO FIND y_s. HENCE

$$y_s = 2 - j2.7 \quad \text{AND} \quad \Gamma_s = 0.71\angle 152^\circ$$

(b)

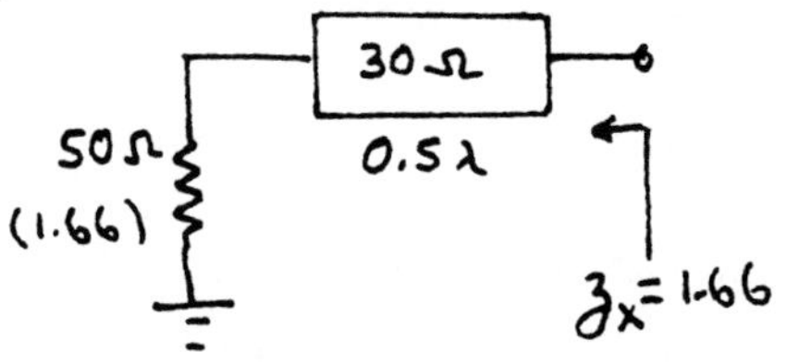

NORMALIZING WITH $Z_0 = 30\Omega$

$$z = \frac{50}{30} = 1.66$$

HENCE, THE IMPEDANCE z_x AT 0.5λ FROM THE LOAD END IS:

$$z_x = 1.66 \text{ OR } Z_x = 30(1.66) = 50\Omega$$

AND $Y_x = \frac{1}{Z_x} = 20 \text{ mS}$

THE 75Ω, 0.46λ BALANCE STUB HAS: $y_{oc} = -j0.26$ OR

$$y_{oc,TOTAL} = 2(-j0.26) = -j0.52$$

$$Y_{oc,TOTAL} = -\frac{j0.52}{75} = -j7 \text{ mS}$$

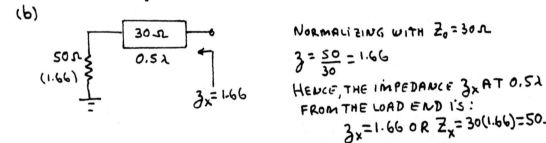

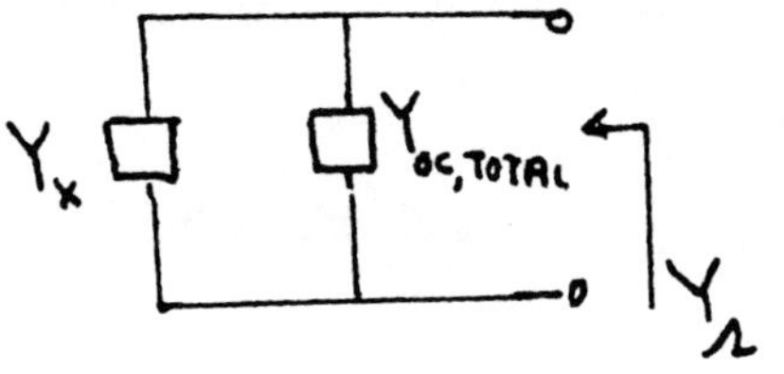

$$Y_s = Y_x + Y_{oc,TOTAL}$$

$$Y_s = 20 - j7 \text{ mS}$$

IN A 50-Ω SYSTEM: $y_s = 50Y_s = 1 - j0.35$

$$z_s = \frac{1}{y_s} = 0.891 + j0.312$$

AND $\Gamma_s = 0.172\angle 99.9^\circ$

2.27) (a) $\lambda = \frac{3\,10^{10}}{6\,10^{9}} = 5\ cm$, $l_1 = 1.25\,cm = 1.25\frac{\lambda}{5} = \frac{\lambda}{4}$

$l_2 = 1.87\,cm = 1.87\frac{\lambda}{5} = 0.375\lambda = \frac{3\lambda}{8}$

FOR THE $\frac{\lambda}{4}$ LINE:

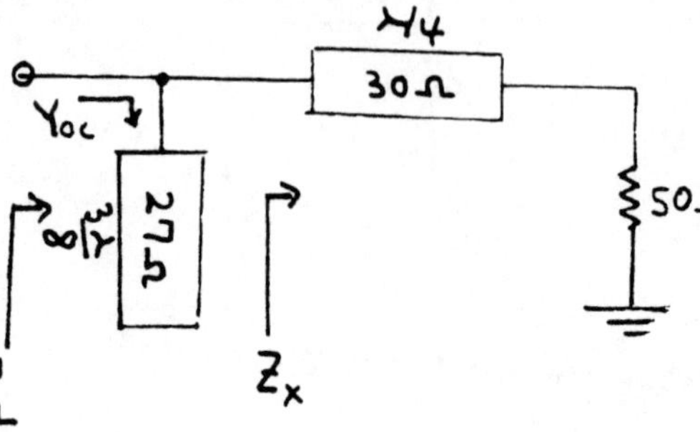

$Z_x = \frac{Z_0^2}{Z_L} = \frac{30^2}{50} = 18\,\Omega$

$Y_x = \frac{1}{Z_x} = 56\ mS$

FOR THE $\frac{3\lambda}{8}$ STUB:

$y_{oc} = -j$, $Y_{oc} = \frac{1}{27}(-j) = -j37\ mS$

HENCE: $Y_L = Y_x + Y_{oc} = 56 - j37\ mS$

IN A 50 Ω SYSTEM: $y_L = 50\,Y_L = 2.8 - j1.85$ AND $\Gamma_L = 0.61\angle 160.2^\circ$

(b) FOR BALANCE STUBS USE $Z_0 = 2(27) = 54\,\Omega$ WITH LENGTHS OF $\frac{3\lambda}{8}$.

2.28) $\Gamma_{IN} = 0.5\angle 100^\circ$, $z_{IN} = 0.527 + j0.692$

THE 50 Ω, 0.15λ TRANSFORMS z_{IN} TO THE IMPEDANCE $z_x = 2.94 - j0.74$,

OR $y_x = \frac{1}{z_x} = 0.32 + j0.08$

THE ADMITTANCE OF THE $\frac{\lambda}{8}$ STUB IS: $y_{oc} = j$. HENCE, FOR THE TWO STUBS: $y_{oc,TOTAL} = j + j = 2j$

THEN, $y_A = y_x + y_{oc,TOTAL} = 0.32 + j0.08 + j2$

$= 0.32 + j2.08$

$Z_A = 50\,z_A = \frac{50}{y_A} = (0.072 - j0.47)50 = 3.6 - j23.5\,\Omega$

2.29) $z_L = \frac{Z_L}{50} = 1 - j$, $z_{IN} = \frac{Z_{IN}}{50} = 0.5 + j0.5$

$y_L = \frac{1}{z_L} = 0.5 + j0.5$, $y_{IN} = \frac{1}{z_{IN}} = 1 - j$

y_L AND y_{IN} ARE ON THE SAME CONSTANT $|\Gamma|$ CIRCLE. HENCE, A SERIES TRANSMISSION LINE OF LENGTH:

$$\ell = 0.338\lambda - 0.088\lambda = 0.25\lambda$$

WILL CHANGE y_L TO y_{IN}.

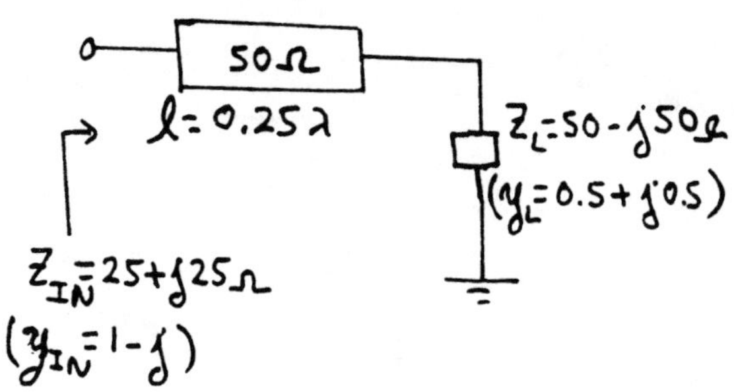

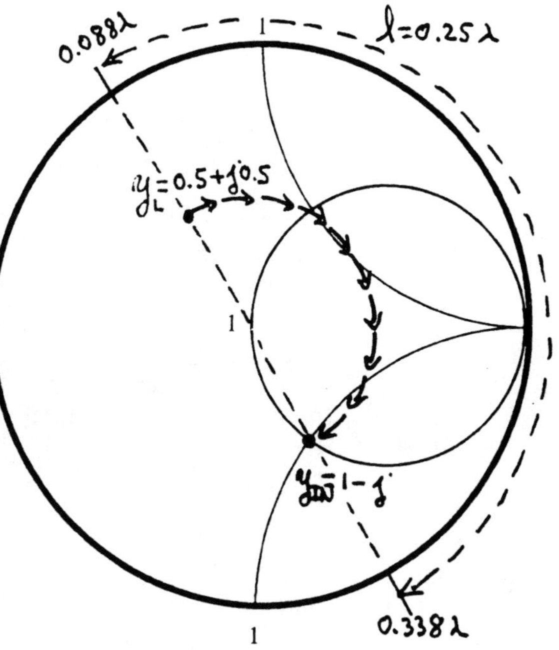

2.30) $z_L = \dfrac{50+j50}{50} = 1+j$

THE 50Ω, 0.125λ CHANGES $z_L = 1+j$ to $z_x = 2-j$ (OR $y_x = 0.4+j0.2$)

THE 50Ω, 0.125λ STUB HAS: $y_{oc} = j$

HENCE: $y_L = y_x + y_{oc} = 0.4 + j0.2 + j = 0.4 + j1.2$ AND $\Gamma_L = 0.728\angle{-104°}$

2.31) LOCATE $\Gamma_\Omega = 0.57\angle 116°$ IN THE SMITH CHART AND READ:

$z_x = 0.27$ OR $Z_x = 50 z_x = 13.5\,\Omega$

THE TRANSFORMATION OF 50Ω to $Z_x = 13.5\,\Omega$ CAN BE DONE WITH A $\frac{\lambda}{4}$ LINE WITH: $Z_o = \sqrt{50(13.5)} = 26\,\Omega$

THEN, A SERIES TRANS. LINE OF LENGTH: $\ell = 0.088\lambda$ TRANSFORMS $Z_x = 13.5\,\Omega$ TO $Z_\Omega = 18.5 + j28\,\Omega$ OR $\Gamma_\Omega = 0.57\angle 116°$

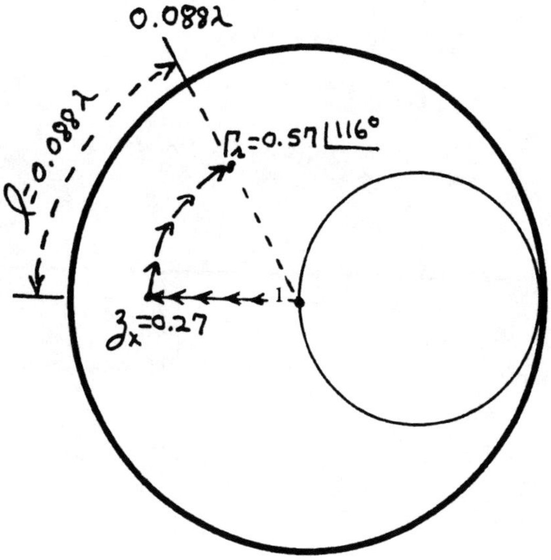

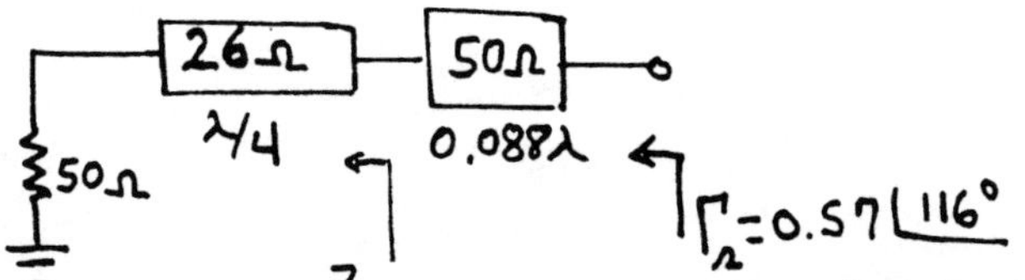

$Z_\Omega = 50(0.37 + j0.562) = 18.5 + j28\,\Omega$

2.32) (a) $z_L = \frac{100+j100}{50} = 2+j2$

THE TRANSMISSION LINE PRODUCES A MOTION ALONG A CONSTANT $|\Gamma|$ CIRCLE, AND THE INDUCTOR L PRODUCES A MOTION ALONG A CONSTANT CONDUCTANCE CIRCLE.

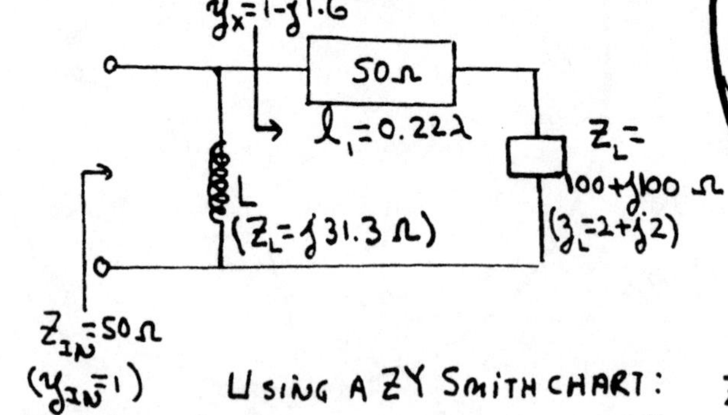

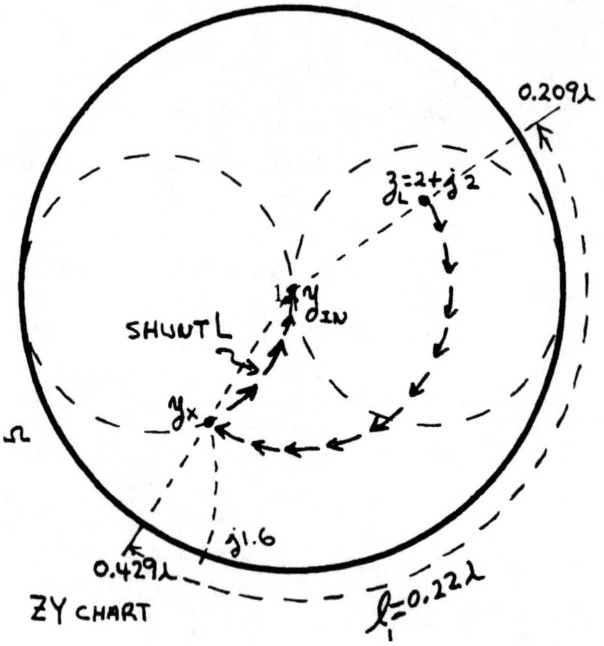

USING A ZY SMITH CHART: ZY CHART

$\ell_1 = 0.429\lambda - 0.209\lambda = 0.22\lambda$

AT y_x: $y_x = 1 - j1.6$. THEN, $y_L = 0 - (j1.6) = -j1.6$

OR $Z_L = 50 z_L = \frac{50}{-j1.6} = j31.3\,\Omega$

(b) IN THIS CIRCUIT THE INDUCTOR PRODUCES A MOTION ALONG A CONSTANT RESISTANCE CIRCLE.

$\ell_1 = 0.322\lambda - 0.209\lambda = 0.113\lambda$

AT z_x: $z_x = 1 - j1.6$

$z_L = 0 - (-j1.6) = j1.6$

$Z_L = 50 z_L = 50(j1.6) = j80\,\Omega$

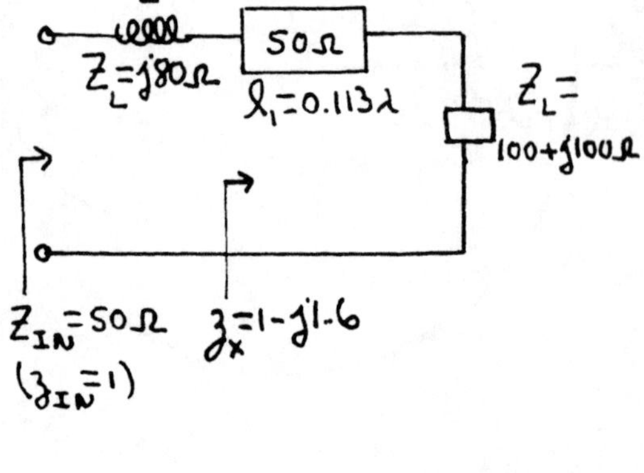

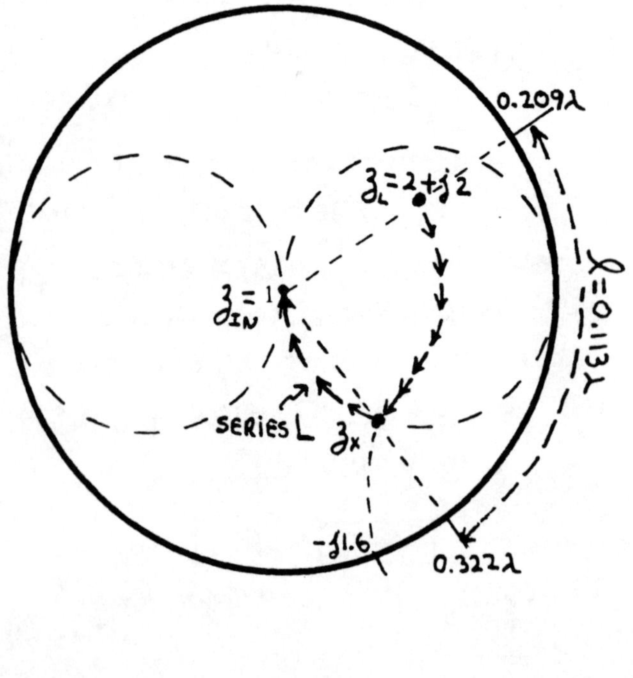

2.33) (a) $z_{IN} = \frac{25 - j25}{50} = 0.5 - j0.5$

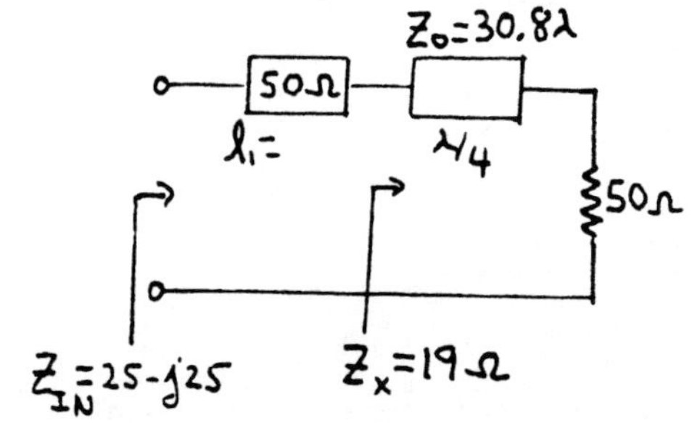

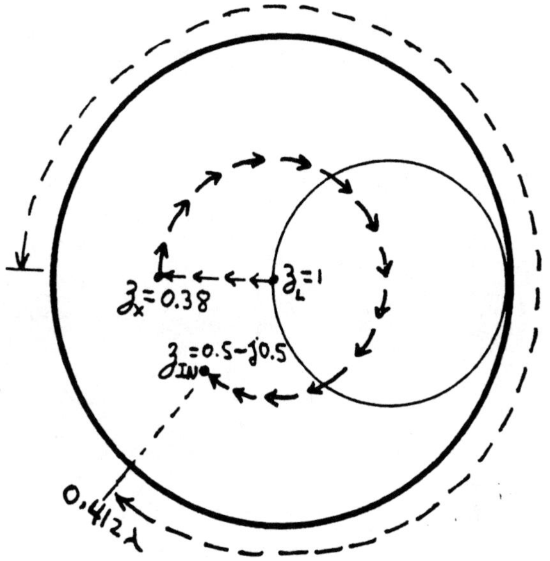

Z_x AND Z_{IN} MUST BE ON THE SAME CONSTANT $|\Gamma|$ CIRCLE. ONE SOLUTION IS SHOWN ON THE SMITH CHART.

$Z_x = 50 z_x = 50(0.38) = 19\Omega$

THEN: $Z_0 = \sqrt{Z_L Z_x} = \sqrt{50(19)} = 30.8\Omega$

AND $l_1 = 0.412\lambda$

(b) $z_{IN} = 0.5 - j0.5$, $y_{IN} = \frac{1}{z_{IN}} = 1 + j$

LETTING $Z_0 = 50\Omega$ AND $l_1 =$ ANY LENGTH, THE ADMITTANCE $y_x = 1$. THEN, THE SHUNT STUB MUST PROVIDE: $y_{s.c} = j$. HENCE: $l_2 = \frac{3\lambda}{8}$.

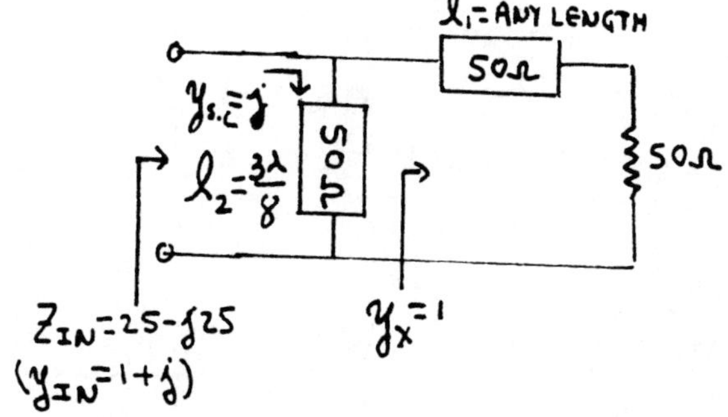

$y_{IN} = y_x + y_{s.c}$
$= 1 + j$

(c) $z_{IN} = \frac{50 + j50}{50} = 1 + j$, $y_{IN} = 0.5 - j0.5$, $Y_{IN} = 10 - j10$ mS

$\therefore$ $Z_0 = \sqrt{Z_L Z_x} = \sqrt{50\left(\frac{1}{10 \cdot 10^{-3}}\right)} = 70.7\Omega$. THE STUB ADMITTANCE MUST BE: $y_{s.c} = -j0.5$. HENCE: $l_2 = 0.426\lambda$

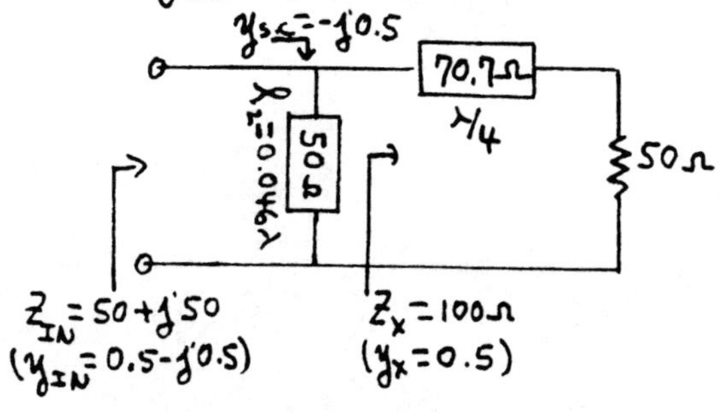

$y_{IN} = y_x + y_{s.c}$
$= 0.5 - j0.5$

2.34) STARTING WITH (2.6.13):

$$G_T = \frac{(1-|\Gamma_s|^2)\,|S_{21}|^2\,(1-|\Gamma_L|^2)}{|(1-S_{11}\Gamma_s)(1-S_{22}\Gamma_L)-S_{21}S_{12}\Gamma_s\Gamma_L|^2} \qquad (1)$$

THE DENOMINATOR CAN BE EXPRESSED AS

$$|1-S_{22}\Gamma_L|^2 \left|1-S_{11}\Gamma_s - \frac{S_{12}S_{21}\Gamma_s\Gamma_L}{1-S_{22}\Gamma_L}\right|^2 =$$

$$|1-S_{22}\Gamma_L|^2 \left|1-\Gamma_s\underbrace{\left(S_{11}-\frac{S_{12}S_{21}\Gamma_L}{1-S_{22}\Gamma_L}\right)}_{\Gamma_{IN}}\right|^2 = |1-S_{22}\Gamma_L|^2\,|1-\Gamma_s\Gamma_{IN}|^2$$

HENCE, (1) CAN BE EXPRESSED IN THE FORM

$$G_T = \frac{1-|\Gamma_s|^2}{|1-\Gamma_s\Gamma_{IN}|^2}\,|S_{21}|^2\,\frac{1-|\Gamma_L|^2}{|1-S_{22}\Gamma_L|^2} \qquad (2.6.14)$$

ANOTHER WAY OF WRITING THE DENOMINATOR IS:

$$|1-S_{11}\Gamma_s|^2 \left|1-S_{22}\Gamma_L - \frac{S_{12}S_{21}\Gamma_s\Gamma_L}{1-S_{11}\Gamma_s}\right|^2 =$$

$$|1-S_{11}\Gamma_s|^2 \left|1-\Gamma_L\underbrace{\left(S_{22}-\frac{S_{12}S_{21}\Gamma_s}{1-S_{11}\Gamma_s}\right)}_{\Gamma_{OUT}}\right|^2 = |1-S_{11}\Gamma_s|^2\,|1-\Gamma_L\Gamma_{OUT}|^2$$

HENCE, (1) CAN ALSO BE EXPRESSED AS:

$$G_T = \frac{1-|\Gamma_s|^2}{|1-S_{11}\Gamma_s|^2}\,|S_{21}|^2\,\frac{1-|\Gamma_L|^2}{|1-\Gamma_L\Gamma_{OUT}|^2} \qquad (2.6.15)$$

2.35) $\frac{b_1}{b_s} = \frac{S_{11}(1-\Gamma_L S_{22})+S_{21}\Gamma_L S_{12}}{D}$, $\frac{b_2}{b_s} = \frac{S_{21}}{D}$

$\frac{a_1}{b_s} = \frac{1-S_{22}\Gamma_L}{D}$, $\frac{a_2}{b_s} = \frac{S_{21}\Gamma_L}{D}$,

$$D = 1-(S_{11}\Gamma_s + S_{22}\Gamma_L + S_{21}\Gamma_L S_{12}\Gamma_s) + S_{11}\Gamma_s S_{22}\Gamma_L$$

HENCE: $$A_v = \frac{\frac{a_2}{b_s}+\frac{b_2}{b_s}}{\frac{a_1}{b_s}+\frac{b_1}{b_s}} = \frac{S_{21}\Gamma_L + S_{21}}{S_{11}(1-\Gamma_L S_{22})+S_{21}\Gamma_L S_{12} + 1 - S_{22}\Gamma_L}$$

OR

$$A_v = \frac{S_{21}(\Gamma_L+1)}{1-S_{22}\Gamma_L + S_{11}(1-\Gamma_L S_{22}) + S_{21}\Gamma_L S_{12}}$$

2.36) (a) FROM (2.8.6), WITH $\Gamma_L = \Gamma_{OUT}^* = 0.682\angle 97^\circ$, WE OBTAIN $|\Gamma_b| = 0$.

THEN, USING (2.8.4), $(VSWR)_{out} = 1$.

(b) WHEN $\Gamma_L = \Gamma_{OUT}^*$ WE HAVE $|\Gamma_b| = 0$ OR $\Gamma_b = 0$. HENCE,

$Z_b = Z_0 = 50\,\Omega$.

(c)
$$|\Gamma_b| = \left| \frac{\Gamma_{OUT} - \Gamma_L^*}{1 - \Gamma_{OUT}\Gamma_L} \right| = \left| \frac{0.5\angle -60^\circ - 0.682\angle -97^\circ}{1 - 0.5\angle -60^\circ (0.682\angle 97^\circ} \right| = 0.546$$

$$(VSWR)_{out} = \frac{1 + 0.546}{1 - 0.546} = 3.41$$

2.37) (a) FROM (2.8.3), WITH $\Gamma_s = \Gamma_{IN}^* = 0.545\angle 77.7^\circ$, WE OBTAIN $|\Gamma_a| = 0$.

THEN, USING (2.8.1), $(VSWR)_{in} = 1$.

(b) WHEN $\Gamma_s = \Gamma_{IN}^*$ WE HAVE $|\Gamma_a| = 0$ OR $\Gamma_a = 0$. HENCE,

$Z_a = Z_0 = 50\,\Omega$

(c)
$$|\Gamma_a| = \left| \frac{\Gamma_{IN} - \Gamma_s^*}{1 - \Gamma_{IN}\Gamma_s} \right| = \left| \frac{0.4\angle 45^\circ - 0.545\angle -77.7^\circ}{1 - 0.4\angle 45^\circ (0.545\angle 77.7^\circ)} \right| = 0.735$$

$$(VSWR)_{in} = \frac{1 + 0.735}{1 - 0.735} = 6.54$$

3.1) (a) $G_T = \dfrac{P_L}{P_{AVS}}$, $G_p = \dfrac{P_L}{P_{IN}}$, $G_A = \dfrac{P_{AVN}}{P_{AVS}}$, $P_{IN} = P_{AVS} M_s \; (M_s \le 1)$,

AND $P_L = P_{AVN} M_L \; (M_L \le 1)$.

HENCE, [SEE (2.7.29)]: $G_T = G_p M_s$ OR $G_T \le G_p$, AND

$G_T = G_A M_L$ OR $G_T \le G_A$.

$G_T = G_p$ WHEN $\Gamma_s = \Gamma_{IN}^*$ (OR WHEN $M_s = 1$)

$G_T = G_A$ WHEN $\Gamma_L = \Gamma_{OUT}^*$ (OR WHEN $M_L = 1$)

(b) G_T IN (3.2.1) AND G_p IN (3.2.3) SHOULD BE IDENTICAL WHEN $\Gamma_s = \Gamma_{IN}^*$ (i.e., $P_{IN} = P_{AVS}$ WHEN $\Gamma_s = \Gamma_{IN}^*$). FROM (3.2.1) WITH $\Gamma_s = \Gamma_{IN}^*$:

$$G_T = \frac{1-|\Gamma_{IN}|^2}{|1-\Gamma_{IN}\Gamma_{IN}^*|^2}|S_{21}|^2\frac{1-|\Gamma_L|^2}{|1-S_{22}\Gamma_L|^2} = \frac{1-|\Gamma_{IN}|^2}{(1-|\Gamma_{IN}|^2)^2}|S_{21}|^2\frac{1-|\Gamma_L|^2}{|1-S_{22}\Gamma_L|^2}$$

HENCE:

$$G_T = G_p = \frac{1}{1-|\Gamma_{IN}|^2}|S_{21}|^2\frac{1-|\Gamma_L|^2}{|1-S_{22}\Gamma_L|^2}$$

G_T IN (3.2.2) AND G_A IN (3.2.4) SHOULD BE IDENTICAL WHEN $\Gamma_L = \Gamma_{OUT}^*$ (i.e, $P_L = P_{AVN}$ WHEN $\Gamma_L = \Gamma_{OUT}^*$). FROM (3.2.2) WITH $\Gamma_L = \Gamma_{OUT}^*$:

$$G_T = \frac{1-|\Gamma_s|^2}{|1-S_{11}\Gamma_s|^2}|S_{21}|^2\frac{1-|\Gamma_{OUT}|^2}{|1-\Gamma_{OUT}\Gamma_{OUT}^*|^2} = \frac{1-|\Gamma_s|^2}{|1-S_{11}\Gamma_s|^2}|S_{21}|^2\frac{1-|\Gamma_{OUT}|^2}{(1-|\Gamma_{OUT}|^2)^2}$$

HENCE:

$$G_T = G_A = \frac{1-|\Gamma_s|^2}{|1-S_{11}\Gamma_s|^2}|S_{21}|^2\frac{1}{1-|\Gamma_{OUT}|^2}$$

3.2) (a) WITH $Z_s = Z_L = Z_0$ THEN $\Gamma_s = \Gamma_L = 0$, AND FROM (3.2.1):

$$G_T = |S_{21}|^2$$

(b) FROM (3.2.5): $\Gamma_{IN} = S_{11}$ WHEN $\Gamma_L = 0$. THEREFORE, FROM (3.2.3):

$$G_p = \frac{|S_{21}|^2}{1-|\Gamma_{IN}|^2} = \frac{|S_{21}|^2}{1-|S_{11}|^2}$$

(c) FROM (3.2.6): $\Gamma_{OUT} = S_{22}$ WHEN $\Gamma_s = 0$. THEREFORE, FROM (3.2.4):

$$G_A = \frac{|S_{21}|^2}{1-|\Gamma_{OUT}|^2} = \frac{|S_{21}|^2}{1-|S_{22}|^2}$$

3.3) (a) WITH THE VALUES GIVEN IN THE PROBLEM IT FOLLOWS THAT:

FROM (3.2.5): $\Gamma_{IN} = 0.671 \angle 160.72^\circ$

FROM (3.2.6): $\Gamma_{OUT} = 0.615 \angle -82.8^\circ$

FROM (3.2.1): $G_T = 8.575$ OR $9.33\,dB$

FROM (3.2.3): $G_p = 9.487$ OR $9.77\,dB$

FROM (3.2.4): $G_A = 8.745$ OR $9.42\,dB$

(b) $P_{AVS} = \dfrac{|E_1|^2}{8\,Re[Z_1]} = \dfrac{10^2}{8(50)} = 0.25\,W$, $M_1 = \dfrac{G_T}{G_p} = \dfrac{8.575}{9.487} = 0.904$

$P_{IN} = P_{AVS} M_1 = 0.25(0.904) = 0.226\,W$

$P_L = G_p P_{IN} = 9.487(0.226) = 2.144\,W$

$P_{AVN} = G_A P_{AVS} = 8.745(0.25) = 2.186\,W$

3.4) $\Gamma_1 = 0$ AND $\Gamma_L = 0.5 \angle 90^\circ$. HENCE:

$G_T = G_{TU} = 11.294$ OR $10.53\,dB$

$G_p = 22.145$ OR $13.45\,dB$

$G_A = 21.33$ OR $13.29\,dB$

3.5) OBSERVE THAT $\Gamma_{OUT} = S_{22}$ WHEN $\Gamma_1 = 0$. THEREFORE, THE ORIGIN (i.e., $\Gamma_1 = 0$) IS A STABLE POINT WHEN $|S_{22}| < 1$.

THE "SHADED" PART IS THE "STABLE REGION".

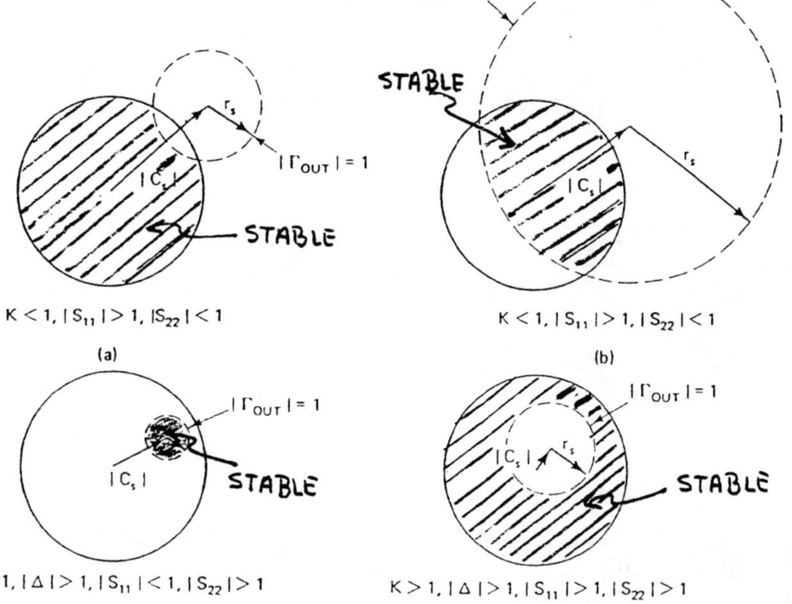

$K < 1, |S_{11}| > 1, |S_{22}| < 1$

(a)

$K < 1, |S_{11}| > 1, |S_{22}| < 1$

(b)

$K > 1, |\Delta| > 1, |S_{11}| < 1, |S_{22}| > 1$

(c)

$K > 1, |\Delta| > 1, |S_{11}| > 1, |S_{22}| > 1$

(d)

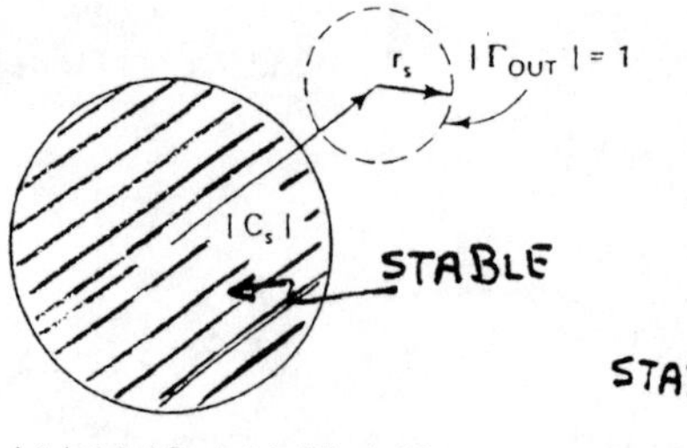

$K > 1, |\Delta| < 1, |S_{11}| < 1, |S_{22}| < 1$

(e)

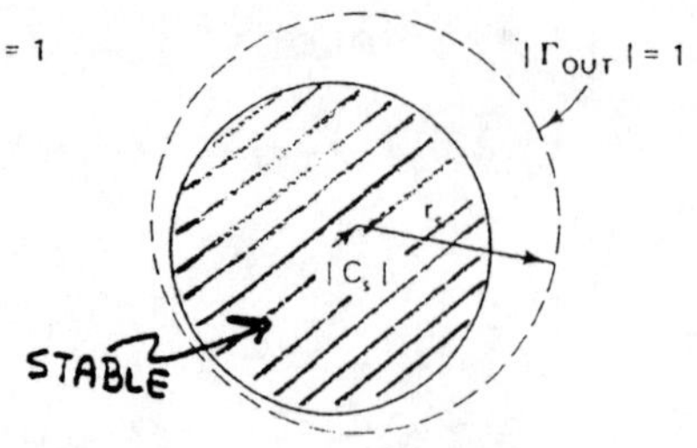

$K > 1, |\Delta| < 1, |S_{11}| < 1, |S_{22}| < 1$

(f)

3.6) OBSERVE THAT $\Gamma_{IN} = S_{11}$ WHEN $\Gamma_L = 0$. THEREFORE, THE ORIGIN (i.e., $\Gamma_L = 0$) IS A STABLE POINT WHEN $|S_{11}| < 1$.

THE "SHADED" PART IS THE "STABLE REGION".

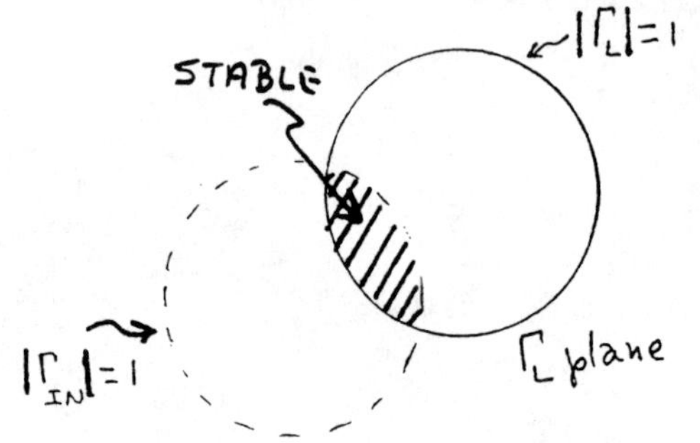

$|S_{11}|>1, |S_{22}|<1$

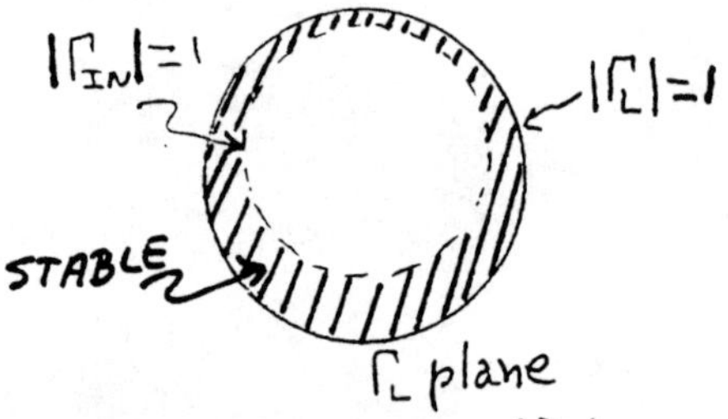

$|S_{11}|>1, |S_{22}|>1$

3.7) (a) $K = 1.284$, $\Delta = 0.386 \angle 134.2°$ ∴ UNCONDITIONALLY STABLE.

(b) $K = 0.909$, $\Delta = 0.402 \angle -65.04°$ ∴ POTENTIALLY UNSTABLE.

INPUT STABILITY CIRCLE $\begin{cases} r_s = 9.27 \\ C_s = 8.37 \angle -57.6° \end{cases}$

OUTPUT STABILITY CIRCLE $\begin{cases} r_L = 0.19 \\ C_L = 1.18 \angle 29.8° \end{cases}$

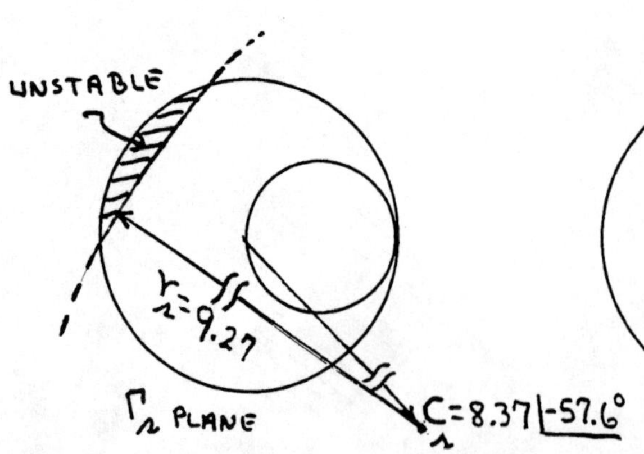

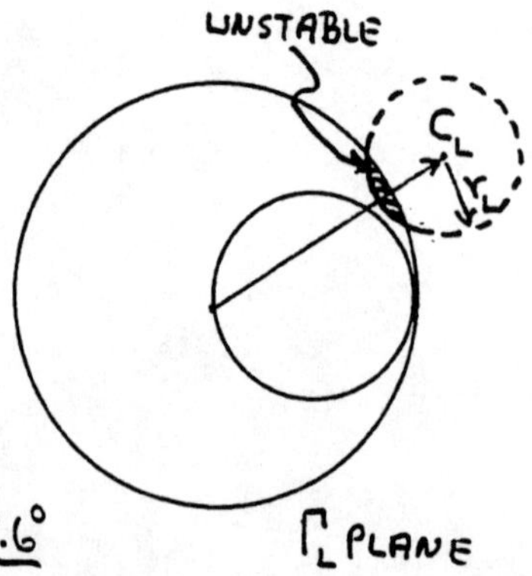

(c) $K = 1.202$, $\Delta = 1.76\angle 18.5°$ ∴ POTENTIALLY UNSTABLE

INPUT STABILITY CIRCLE: $r_s = 0.518$, $C_s = 0.152\angle 82.1°$

OUTPUT STABILITY CIRCLE: $r_L = 0.494$, $C_L = 0.239\angle -58°$

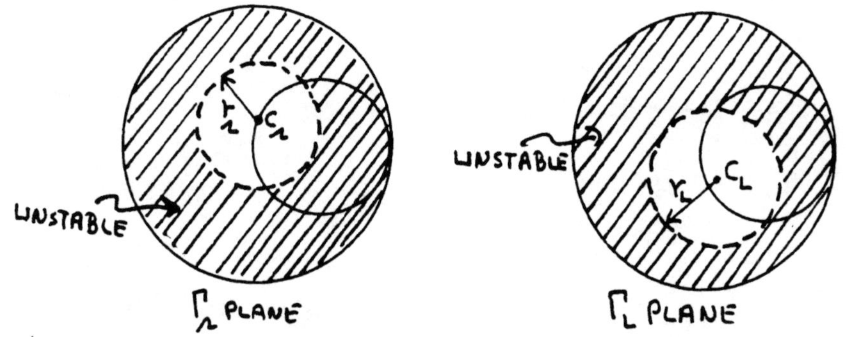

3.8)

f(GHz)	K	Δ	STABILITY CIRCLE	STABILITY CIRCLE
4	0.412	$0.65\angle -90.3°$	$r_s = 0.444$, $C_s = 1.25\angle 80.9°$	$r_L = 3.79$, $C_L = 4.3\angle 91.4°$
6	0.56	$0.57\angle -131.1°$	$r_s = 0.604$, $C_s = 1.43\angle 111.6°$	$r_L = 6.316$, $C_L = 6.93\angle 107.8°$
8	0.78	$0.43\angle 174.9°$	$r_s = 0.789$, $C_s = 1.69\angle 152.6°$	$r_L = 7.39$, $C_L = 8.2\angle 126.1°$
10	0.89	$0.32\angle 114°$	$r_s = 0.759$, $C_s = 1.72\angle -171.8°$	$r_L = 3.49$, $C_L = 4.41\angle 150.1°$
14	1.33	$0.17\angle -2.4°$	$r_s = 0.513$, $C_s = 1.62\angle -110.2°$	$r_L = 1.87$, $C_L = 3.08\angle -171.7°$

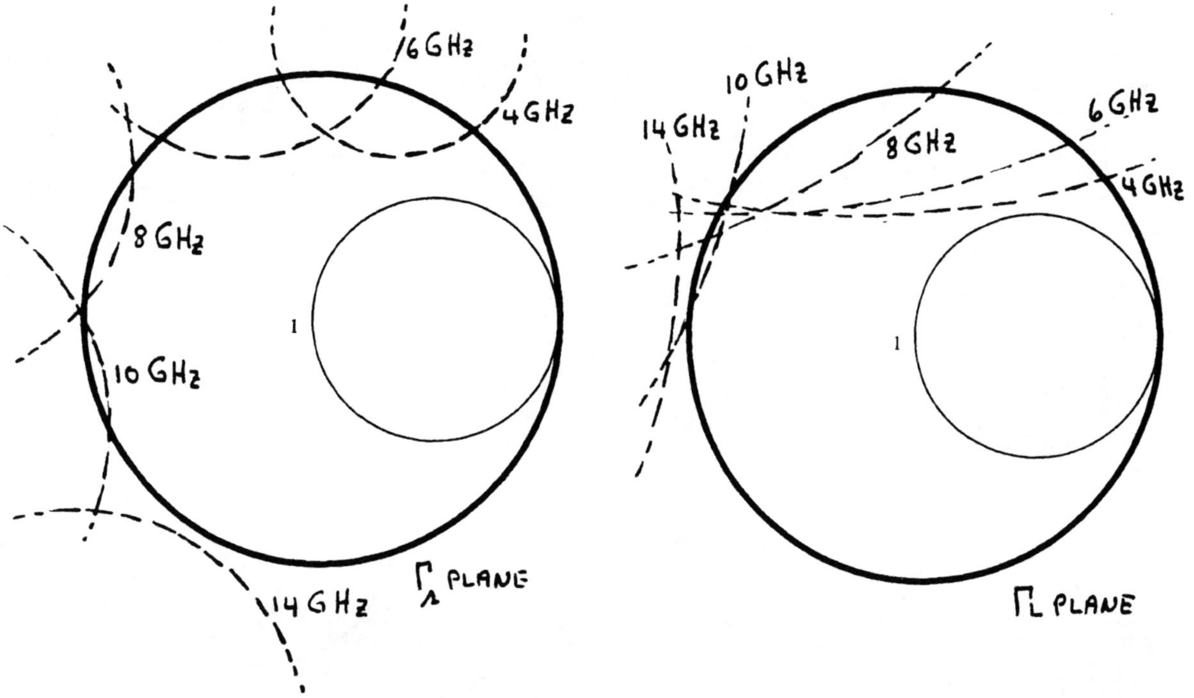

3.9) (a) $K=1$, $\Delta=1$

INPUT STABILITY CIRCLE:

$$\begin{cases} r_s = 1 \\ C_s = 0 \end{cases}$$

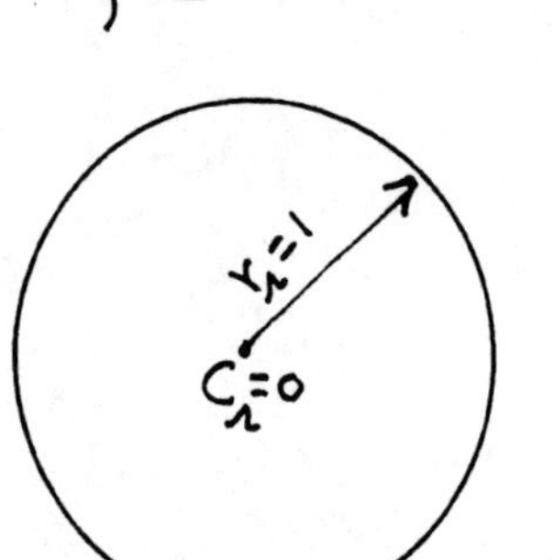

OUTPUT STABILITY CIRCLE:

$$\begin{cases} r_L = 1 \\ C_L = 0 \end{cases}$$

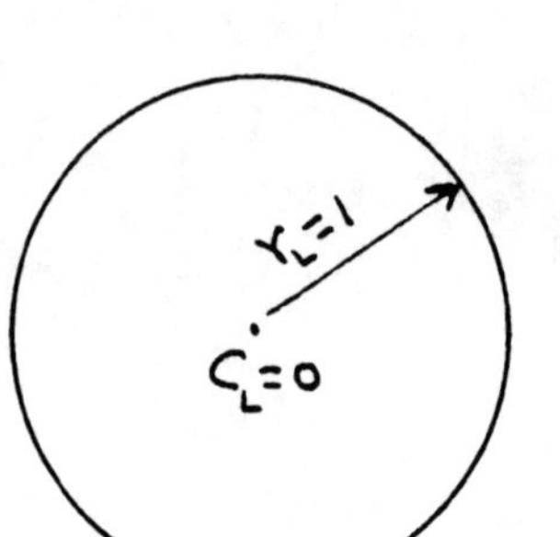

(b) $K=1$, $\Delta=-2.414$

$$\begin{cases} r_s = 0.55 \\ C_s = 0.45\angle 180^\circ \end{cases}$$

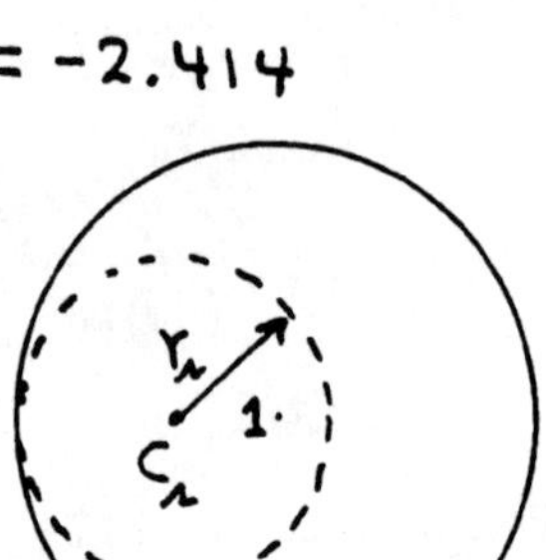

$$\begin{cases} r_L = 0.55 \\ C_L = 0.45\angle 180^\circ \end{cases}$$

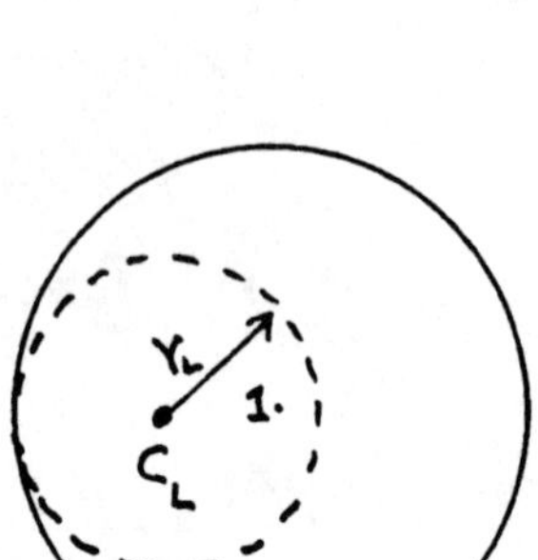

(c) $K=1$, $\Delta=0.415$

$$\begin{cases} r_s = 0.26 \\ C_s = 1.26 \end{cases}$$

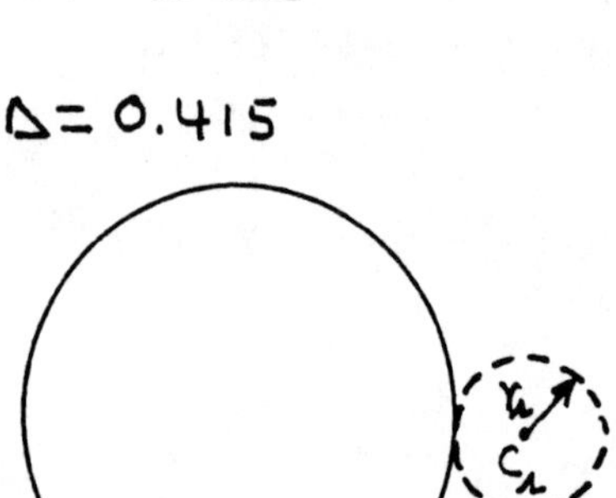

$$\begin{cases} r_L = 0.26 \\ C_L = 1.26 \end{cases}$$

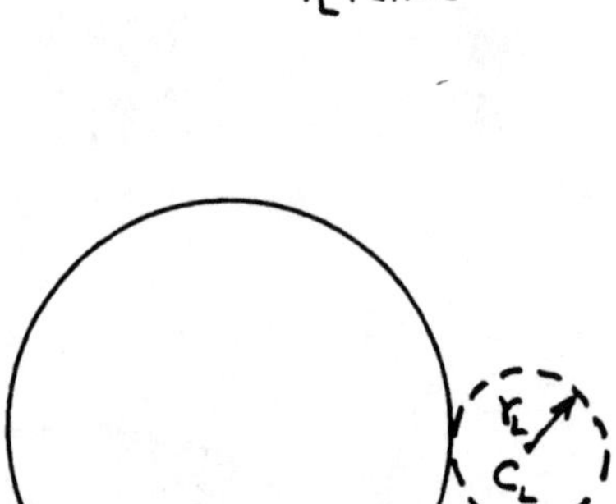

(d) $K=0$, $\Delta=-1$

$$\begin{cases} r_s = \infty \\ C_s = \infty \end{cases}$$

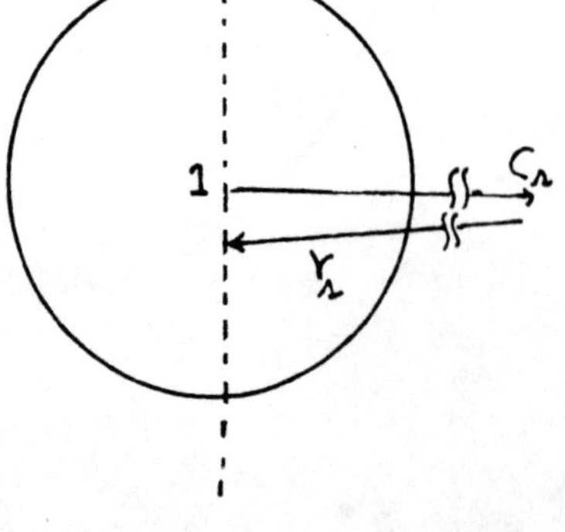

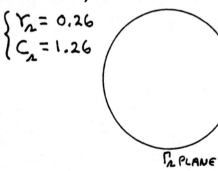

$$\begin{cases} r_L = \infty \\ C_L = \infty \end{cases}$$

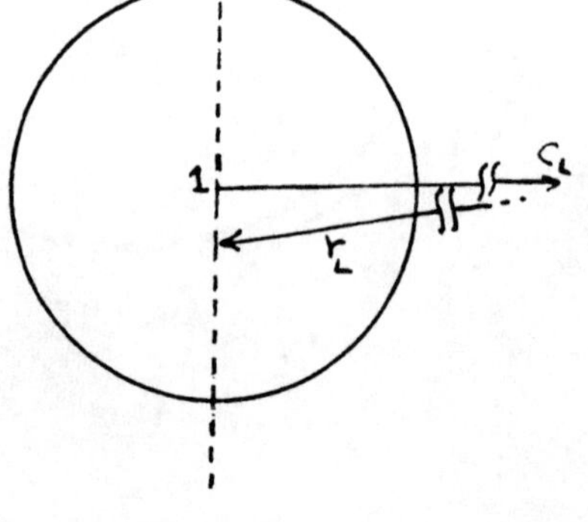

3.10) (a) FROM (3.3.7) AND (3.3.8): $\lim_{S_{12}\to 0} \Delta = S_{11}S_{22}$. THEN,

$$\lim_{S_{12}\to 0} r_L = 0$$

AND

$$\lim_{S_{12}\to 0} C_L = \frac{(S_{22} - S_{11}S_{22}S_{11})^*}{|S_{22}|^2 - |S_{11}S_{22}|^2} = \frac{S_{22}^*(1-|S_{11}|^2)}{|S_{22}|^2(1-|S_{11}|^2)} = \frac{S_{22}^*}{|S_{22}|^2} = \frac{1}{S_{22}}$$

SIMILARLY: $\lim_{S_{12}\to 0} r_s = 0$ AND $\lim_{S_{12}\to 0} C_s = \frac{1}{S_{11}}$

(b) $\begin{cases} r_s = 0 \\ C_s = \frac{1}{S_{11}} = 0.5\angle -90^\circ \end{cases}$ $\quad\quad$ $\begin{cases} r_L = 0 \\ C_L = \frac{1}{S_{22}} = 10\angle -45^\circ \end{cases}$

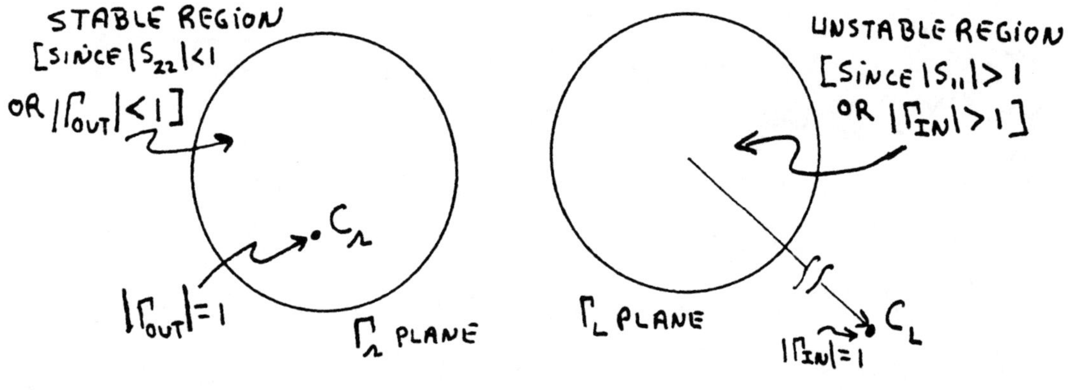

3.11) THE SOURCE STABILITY CIRCLE DOES NOT ENCLOSE THE CENTER OF THE SMITH CHART WHEN: $|C_s| > r_s$. FROM (3.3.9) AND (3.3.10) WE OBTAIN: $|S_{11} - \Delta S_{22}^*| > |S_{12}S_{21}|$

FROM PROBLEM 3.20: $|S_{11} - \Delta S_{22}^*|^2 = |S_{12}S_{21}|^2 + (1-|S_{22}|^2)(|S_{11}|^2 - |\Delta|^2)$

HENCE: $|S_{12}S_{21}|^2 + (1-|S_{22}|^2)(|S_{11}|^2-|\Delta|^2) > |S_{12}S_{21}|^2$

OR $\quad (1-|S_{22}|^2)(|S_{11}|^2 - |\Delta|^2) > 0 \quad (1)$

FROM (1): $\begin{cases} \text{If } |S_{22}|<1 \text{ then } |\Delta| < |S_{11}| \\ \text{If } |S_{22}|>1 \text{ then } |\Delta| > |S_{11}| \end{cases}$

SIMILARLY: $|C_L| > r_L \Rightarrow |S_{22} - \Delta S_{11}^*| > |S_{12}S_{21}|$ OR

$(1-|S_{11}|^2)(|S_{22}|^2 - |\Delta|^2) > 0 \quad (2)$

FROM (2): $\begin{cases} \text{If } |S_{11}|<1 \text{ then } |\Delta| < |S_{22}| \\ \text{If } |S_{11}|>1 \text{ then } |\Delta| > |S_{22}| \end{cases}$

3.12) THE ANSWER TO THIS PROBLEM FOLLOWS FROM THE RESULTS IN APPENDIX J. THAT IS, WITH $C_{ii}=0$ AND $r_{ii}=1$ WE OBTAIN:

$$C_{OUT}=\frac{S_{22}-\Delta S_{11}^*}{1-|S_{11}|^2} \quad \text{AND} \quad r_{OUT}=\frac{|S_{12}S_{21}|}{1-|S_{11}|^2}$$

ALSO, WITH $C_{oo}=0$ AND $r_{oo}=1$ WE OBTAIN:

$$C_{IN}=\frac{S_{11}-\Delta S_{22}^*}{1-|S_{22}|^2} \quad \text{AND} \quad r_{IN}=\frac{|S_{12}S_{21}|}{1-|S_{22}|^2}$$

3.13) (a) $\Delta=S_{11}S_{22}-S_{12}S_{21}$ AND $S_{11}S_{22}=\Delta+S_{12}S_{21}$

$\therefore$ $|\Delta|=|S_{11}S_{22}-S_{12}S_{21}|$ AND $|S_{11}S_{22}|=|\Delta+S_{12}S_{21}|$

HENCE: $|\Delta|\le|S_{11}S_{22}|+|S_{12}S_{21}|$ (1) $\quad |S_{11}S_{22}|\le|\Delta|+|S_{12}S_{21}|$ (2)

FROM (3.3.13): $K>1$ OR $1-|S_{11}|^2-|S_{22}|^2+|\Delta|^2>2|S_{12}S_{21}|$ (3)

(1) INTO (3): $1-|S_{11}|^2-|S_{22}|^2+|\Delta|^2>2|\Delta|-2|S_{11}S_{22}|$ (4)

(2) INTO (3): $1-|S_{11}|^2-|S_{22}|^2+|\Delta|^2>2|S_{11}S_{22}|-2|\Delta|$ (5)

FROM (4): $(1-|\Delta|)^2>(|S_{11}|-|S_{22}|)^2$ (6)

FROM (5): $(1+|\Delta|)^2>(|S_{11}|+|S_{22}|)^2$ (7)

FROM (6) AND (7): $(1-|\Delta|)^2(1+|\Delta|)^2>(|S_{11}|-|S_{22}|)^2(|S_{11}|+|S_{22}|)^2$

$$(1-|\Delta|^2)^2>(|S_{11}|^2-|S_{22}|^2)^2 \quad (8)$$

SINCE: $B_1=1+|S_{11}|^2-|S_{22}|^2-|\Delta|^2$ AND $B_2=1+|S_{22}|^2-|S_{11}|^2-|\Delta|^2$

THEN: $B_1B_2=(1-|\Delta|^2)^2-(|S_{11}|^2-|S_{22}|^2)^2$ (9), $B_1+B_2=2(1-|\Delta|^2)$ (10)

FROM (8) AND (9) WE HAVE: $B_1B_2>0$ (11)

(b) THE RELATION $B_1B_2>0$ SHOWS THAT IF $B_1>0$ THEN $B_2>0$. THEREFORE, THE CONDITIONS $K>1$ AND $B_1>0$ ARE SIMILAR TO $K>1$ AND $B_2>0$.

(c) IF $B_1>0$ THEN $B_2>0$. HENCE, FROM (10):

$$2(1-|\Delta|^2)>0$$

OR $\quad |\Delta|<1$

3.14) FOR THIS TRANSISTOR: $K = 0.532$, $\Delta = 0.617\angle{-85.4^\circ}$

$r_s = 1.96$ $\quad$ $r_L = 0.576$

$C_s = 2.64\angle{116.7^\circ}$ $\quad$ $C_L = 1.4\angle{41.5^\circ}$

AT POINT A: $r = 0.25$

$\therefore R = 0.25(50) = 12.5\,\Omega$

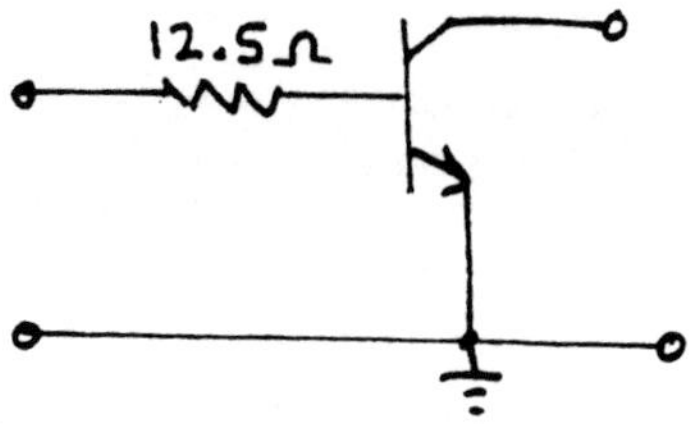

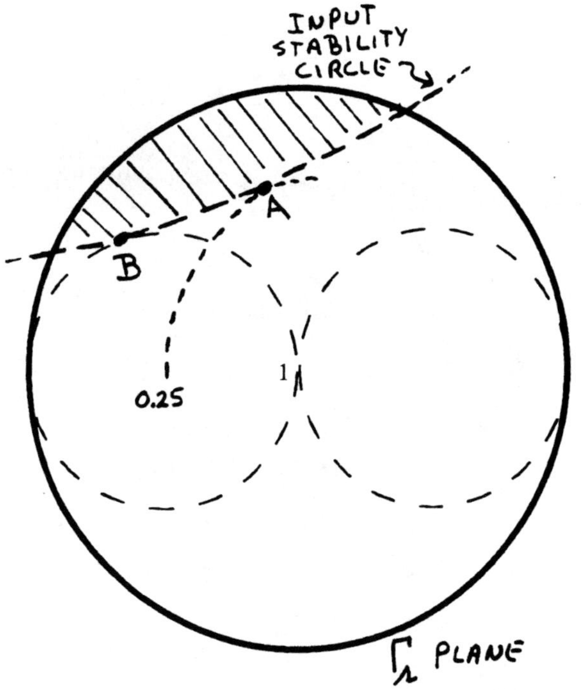

AT POINT B: $g = 1$

$\therefore G = \frac{1}{50} = 20\,mS$ (OR $50\,\Omega$)

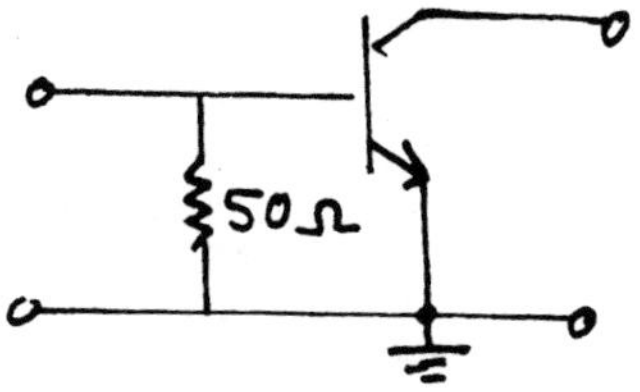

AT POINT C: $r = 1$

$\therefore R = 1(50) = 50\,\Omega$

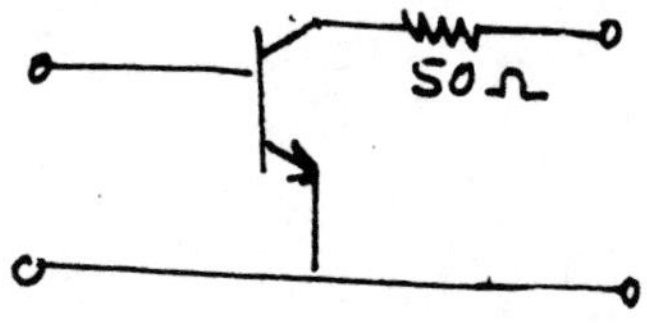

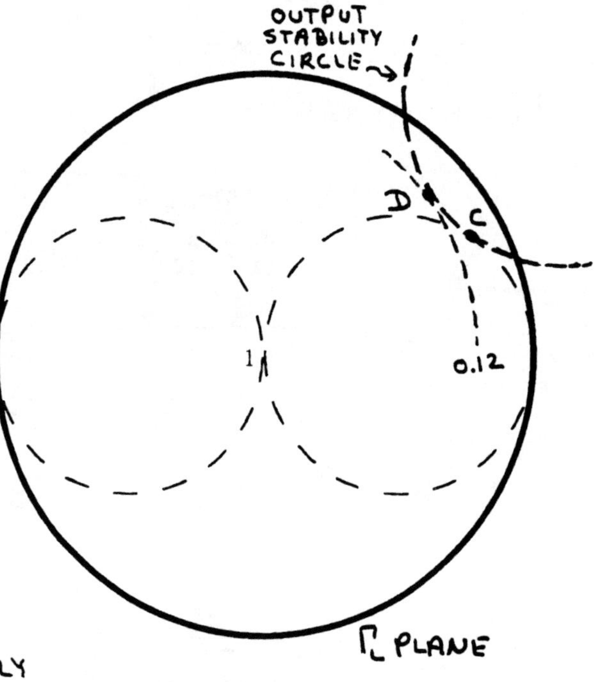

AT POINT D: $g = 0.12$

$G = \frac{0.12}{50} = 2.4\,mS$ (OR $417\,\Omega$)

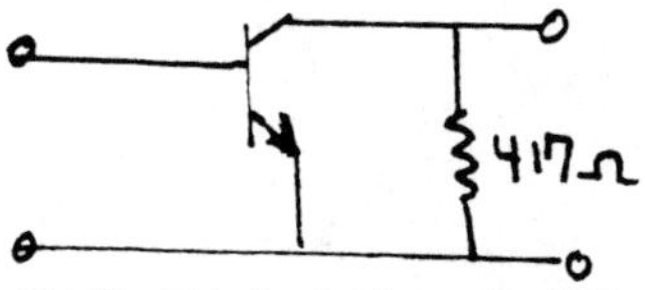

FOR THIS CIRCUIT IT FOLLOWS THAT: $S_{11} = 0.695\angle{-77.3^\circ}$, $S_{12} = 0.03\angle{42.5^\circ}$, $S_{21} = 5.13\angle{124.1^\circ}$, AND $S_{22} = 0.665\angle{-25.8^\circ}$.

HENCE, $K = 1.02$ AND $\Delta = 0.488\angle{-84.7^\circ}$

AND THE CIRCUIT IS UNCONDITIONALLY STABLE.

3.15) THE MAXIMUM VALUE OF $G_i = \dfrac{1-|\Gamma_i|^2}{|1-S_{ii}\Gamma_i|^2}$ IS OBTAINED AS FOLLOWS:

LET $\Gamma_i = x + jy$ AND $S_{ii} = a + jb$. THEN

$$G_i = \frac{1-x^2-y^2}{(1-xa+yb)^2+(xb+ya)^2} = \frac{1-x^2-y^2}{D}\ ;\ D=(1-xa+yb)^2+(xb+ya)^2$$

$$\begin{cases} \dfrac{\partial G_i}{\partial x} = \dfrac{-2x}{D} - \dfrac{(1-x^2-y^2)\left[-2a(1-xa+yb)+2b(xb+ya)\right]}{D^2} = 0 \\[2ex] \dfrac{\partial G_i}{\partial y} = \dfrac{-2y}{D} - \dfrac{(1-x^2-y^2)\left[2b(1-xa+yb)+2a(xb+ya)\right]}{D^2} = 0 \end{cases}$$

OR

$$\begin{cases} 2xD = -(1-x^2-y^2)\left[-2a(1-xa+yb)+2b(xb+ya)\right] & (1) \\ 2yD = -(1-x^2-y^2)\left[2b(1-xa+yb)+2a(xb+ya)\right] & (2) \end{cases}$$

DIVIDING (1) BY (2) GIVES:

$$\frac{x}{y} = \frac{-a(1-xa+yb)+b(xb+ya)}{b(1-xa+yb)+a(xb+ya)} = \frac{-a+xa^2+xb^2}{b+yb^2+ya^2}$$

OR

$$xb + yb^2x + ya^2x = -ay + xa^2y + xb^2y \Rightarrow x = -\frac{ay}{b} \quad (3)$$

(3) INTO (1): $-2\dfrac{ay}{b}\left(1+\dfrac{a^2y}{b}+yb\right)^2 = -\left(1-\dfrac{a^2y^2}{b^2}-y^2\right)\left(-2a\left(1+\dfrac{a^2y}{b}+yb\right)\right)$

OR

$$-\frac{y}{b}\left(1+\frac{a^2y}{b}+yb\right) = 1-\frac{a^2y^2}{b^2}-y^2 \Rightarrow y=-b \quad (4)$$

(4) INTO (3): $x = -\dfrac{a}{b}(-b)$ OR $x = a$ (5)

FROM (4) AND 5: $\Gamma_i = x + jy = a - jb$ OR $\Gamma_i = S_{ii}^*$.

3.16) (a) FOR $G_{TU,max}$: $\Gamma_s = S_{11}^* = 0.706\angle 160^\circ$ AND

$\Gamma_L = S_{22}^* = 0.508\angle 20^\circ$.

$$G_{s,max} = \frac{1}{1-|S_{11}|^2} = \frac{1}{1-(0.706)^2} = 1.99 \text{ OR } 3\text{ dB}$$

$$G_o = |S_{21}|^2 = (5.01)^2 = 25.1 \text{ OR } 14\text{ dB}$$

$$G_{L,max} = \frac{1}{1-|S_{22}|^2} = \frac{1}{1-(0.508)^2} = 1.35 \text{ OR } 1.3\text{ dB}$$

$$G_{TU,max} = 3+14+1.3 = 18.3\text{ dB}$$

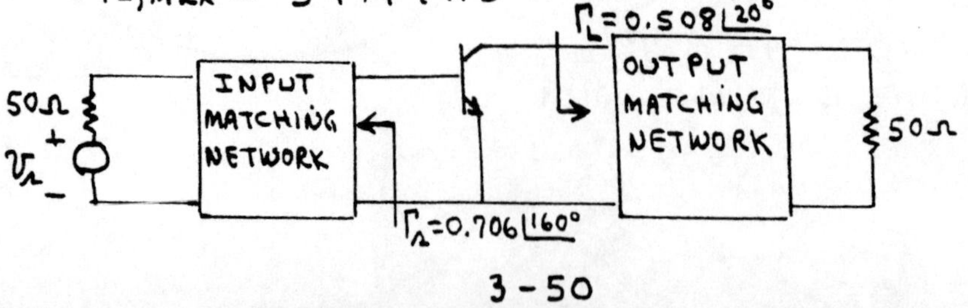

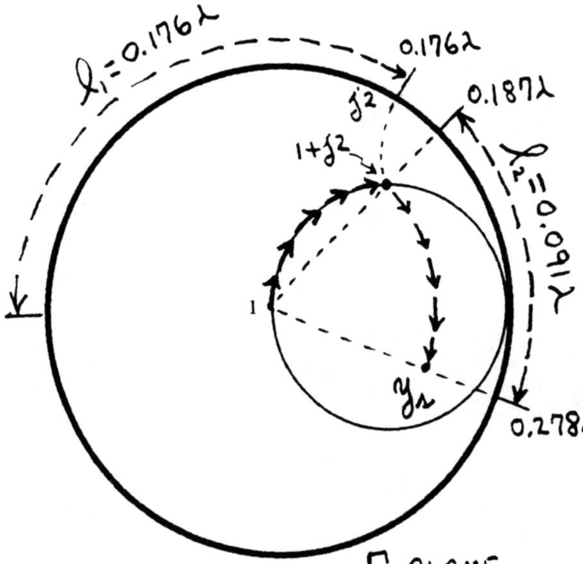

Γ_s PLANE

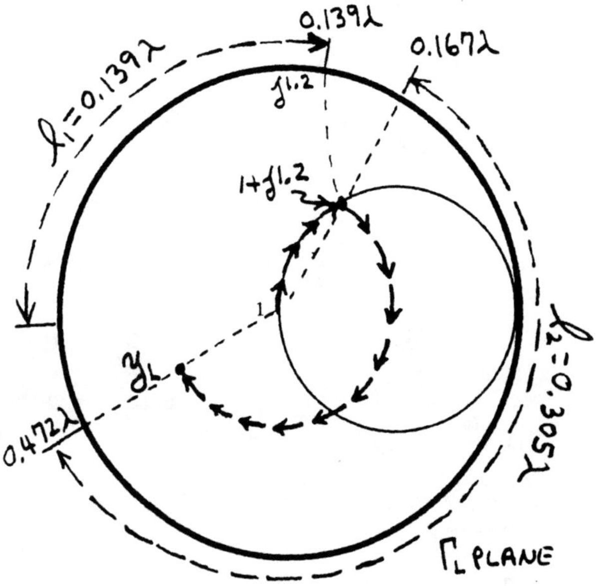

Γ_L PLANE

$\Gamma_s = 0.706\angle 160°$, $y_s = \dfrac{1}{z_s} = 2.9 - j2.8$

$\Gamma_L = 0.508\angle 20°$, $y_L = \dfrac{1}{z_L} = 0.335 - j0.157$

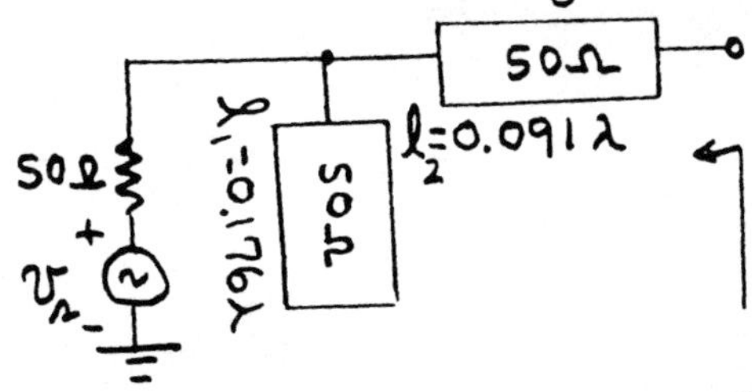

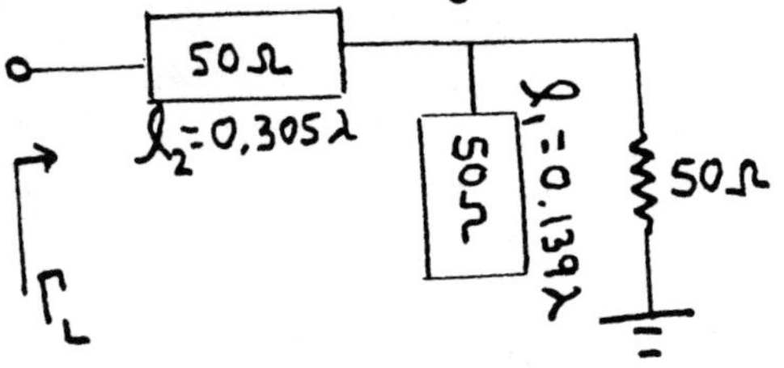

(b)

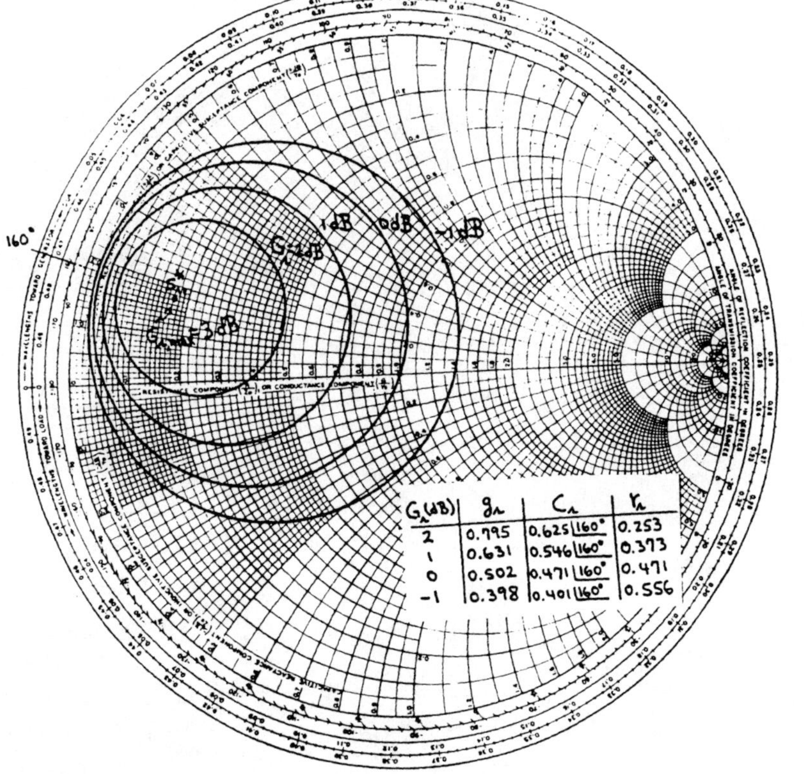

G_s(dB)	g_s	C_s	r_s
2	0.795	0.625∠160°	0.253
1	0.631	0.546∠160°	0.373
0	0.502	0.471∠160°	0.471
−1	0.398	0.401∠160°	0.556

3.17) (a) THE INPUT RESISTANCE IS CALCULATED AS FOLLOWS:

$$\Gamma_{IN} = S_{11} = 2.3\angle{-135^\circ}, \quad Z_{IN} = 50(-0.45 - j0.341) = -22.48 - j17.04\ \Omega.$$

Z_{IN} CAN ALSO BE CALCULATED USING THE SMITH CHART.

PLOT $\frac{1}{S_{11}^*} = 0.435\angle{-135}$ AND READ $Z_{IN} = 50(-0.45 - j0.34) = -22.5 - j17\ \Omega$

FOR $G_s = 4$ dB, $g_s = 2.512[1 - (2.3)^2] = -10.78$

FROM (3.4.11) AND (3.4.12): $C_s = 0.404\angle{135^\circ}$ AND $r_s = 0.24$

(b) AT POINT A, Γ_s HAS THE LARGEST REAL PART ON THE $G_s = 4$ dB CIRCLE. THAT IS,

$\Gamma_s = 0.2\angle{115^\circ}$

THE INPUT MATCHING CIRCUIT MUST TRANSFORM 50 Ω TO $\Gamma_s = 0.2\angle{115^\circ}$.

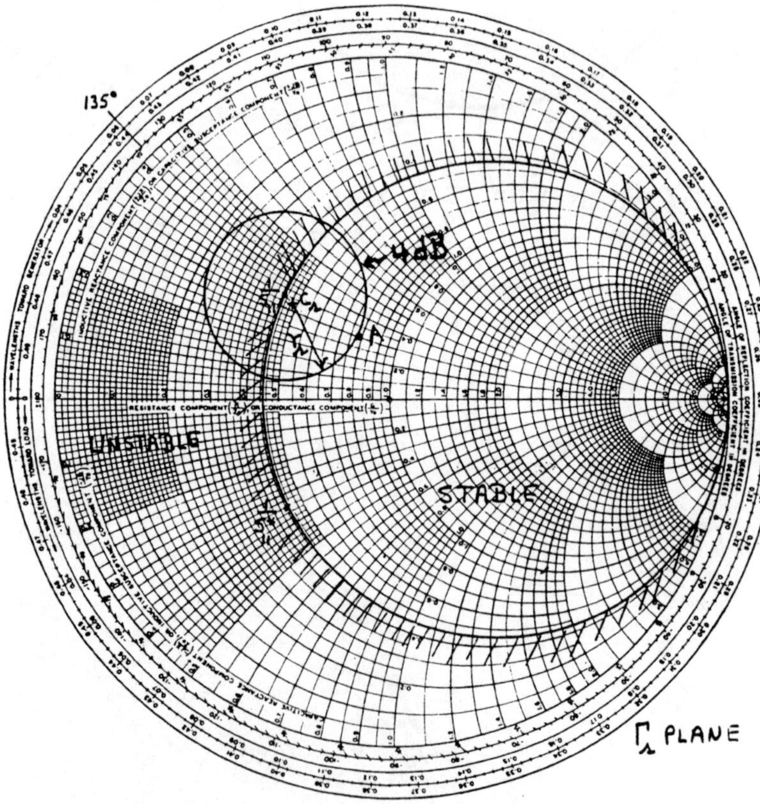

(c) DESIGN FOR $\Gamma_s = 0.2\angle{115^\circ}$ AND $\Gamma_L = S_{22}^* = 0.8\angle{66^\circ}$. THEN,

$G_s = 4$ dB, $G_o = |S_{21}|^2 = 16$ (OR 12.04 dB), $G_{L,max} = \frac{1}{1-(0.8)^2} = 2.78$ (OR 4.4 dB)

$$\therefore\ G_{TU}(\text{dB}) = G_s + G_o + G_{L,max} = 4 + 12.04 + 4.4 = 20.4\ \text{dB}$$

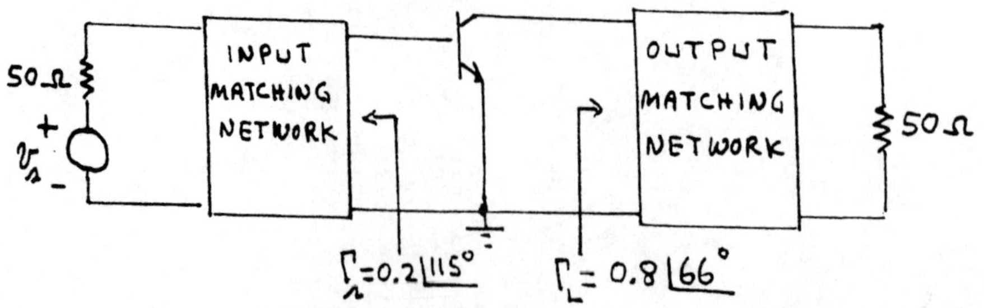

3.18) (a) $G_{s,max}=\frac{1}{1-(0.5)^2}=1.33$ OR $1.25\,dB$, $G_{L,max}=\frac{1}{1-(0.6)^2}=1.563$ OR $1.94\,dB$

$G_o=|S_{21}|^2=25$ OR $13.98\,dB$, $\therefore G_{TU,max}=1.25+13.98+1.94=17.2\,dB$

(b) A MATCHING NETWORK DESIGN AT 900 MHz IS:

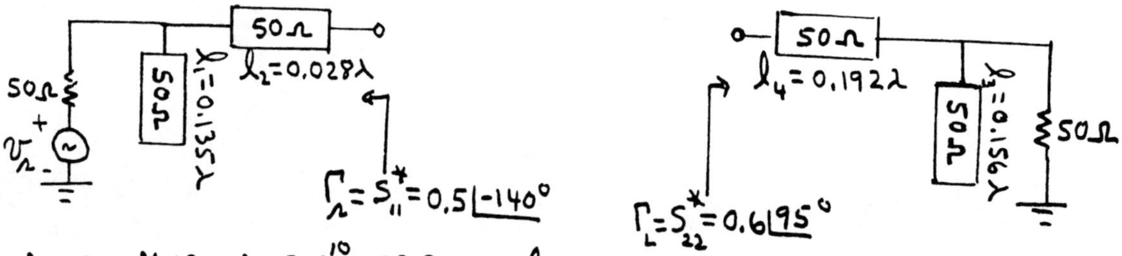

AT 900 MHz: $\lambda=\frac{3\,10^{10}}{9\,10^8}=33.3\,cm$, $l_1=0.135\lambda=4.496\,cm$,

$l_2=0.028\lambda=0.932\,cm$, $l_3=0.156\lambda=5.195\,cm$, $l_4=0.192\lambda=6.394\,cm$

(c) $g_L=\frac{G_L}{G_{L,max}}=\frac{1.259}{1.563}=0.805$

FROM (3.4.11) AND (3.4.12):

$C_L=0.519\angle 95°$

$r_L=0.304$

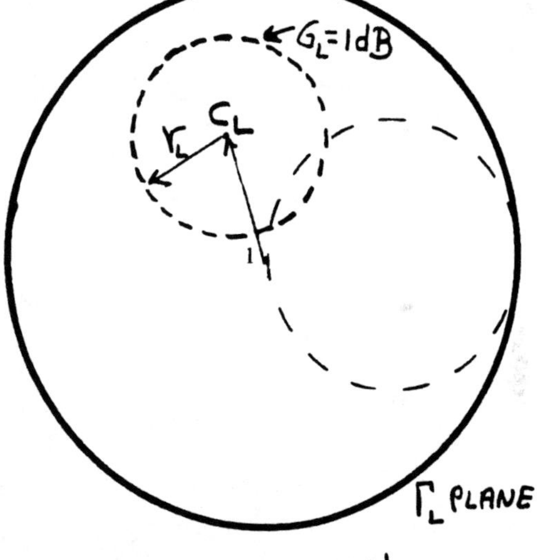

(d) LET $\lambda'=\frac{c}{f'}$ WHERE $f'=1\,GHz$,

AND $\lambda=\frac{c}{f}$ WHERE $f=900\,MHz$.

$\therefore \frac{\lambda}{\lambda'}=\frac{f'}{f}$ OR $\lambda=\frac{f'}{f}\lambda'=\frac{10^9}{9\,10^8}\lambda'=1.11\lambda'$

$l_1=0.135\lambda=0.135(1.11\lambda')=0.15\lambda'$, $l_2=0.028(1.11\lambda')=0.031\lambda'$,

$l_3=0.156(1.11\lambda')=0.173\lambda'$, $l_4=0.192(1.11\lambda')=0.213\lambda'$

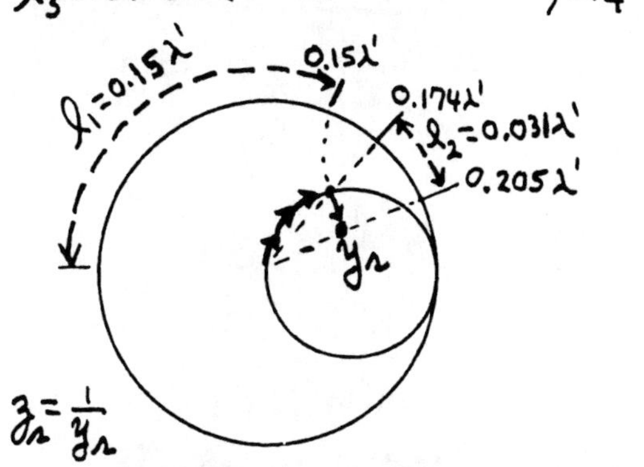

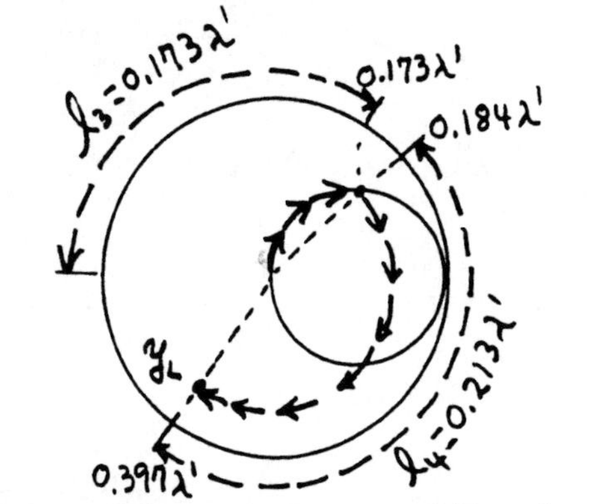

$y_s=1.79+j1.58$, $\Gamma_s=0.55\angle{-146°}$

$y_L=0.28-j0.72$, $\Gamma_L=0.693\angle 74.4°$

$\therefore G_{TU}=\frac{1-(0.55)^2}{|1-0.48\angle 137°(0.55\angle{-146°})|^2}(4.6)^2\frac{1-(0.693)^2}{|1-0.57\angle{-99°}(0.693\angle 74.4°)|^2}=31.97$ OR $15.05\,dB$

3.19) (a) $\Gamma_{MS}^* = S_{11} + \dfrac{S_{12}S_{21}\Gamma_{ML}}{1 - S_{22}\Gamma_{ML}} = \dfrac{S_{11} - \Delta\Gamma_{ML}}{1 - S_{22}\Gamma_{ML}}$ (1)

$\Gamma_{ML}^* = S_{22} + \dfrac{S_{12}S_{21}\Gamma_{MS}}{1 - S_{11}\Gamma_{MS}} = \dfrac{S_{22} - \Delta\Gamma_{MS}}{1 - S_{11}\Gamma_{MS}}$ (2)

SOLVING (2) FOR Γ_{MS} GIVES: $\Gamma_{MS} = \dfrac{\Gamma_{ML}^* - S_{22}}{S_{11}\Gamma_{ML}^* - \Delta}$ (3)

EQUATING (1) AND (3): $\dfrac{S_{11} - \Delta\Gamma_{ML}}{1 - S_{22}\Gamma_{ML}} = \dfrac{\Gamma_{ML} - S_{22}^*}{S_{11}^*\Gamma_{ML} - \Delta^*}$

$$|S_{11}|^2\Gamma_{ML} - S_{11}\Delta^* - \Delta S_{11}^*\Gamma_{ML}^2 + |\Delta|^2\Gamma_{ML} = \Gamma_{ML} - S_{22}^* - S_{22}\Gamma_{ML}^2 + |S_{22}|^2\Gamma_{ML}$$

$$\Gamma_{ML}^2(S_{22} - \Delta S_{11}^*) - \Gamma_{ML}(1 + |S_{22}|^2 - |S_{11}|^2 - |\Delta|^2) + S_{22}^* - S_{11}\Delta^* = 0$$

OR

$$\Gamma_{ML}^2 - \Gamma_{ML}\frac{B_2}{C_2} + \frac{C_2^*}{C_2} = 0 \quad \text{WHERE} \begin{cases} B_2 = 1 + |S_{22}|^2 - |S_{11}|^2 - |\Delta|^2 \\ C_2 = S_{22} - \Delta S_{11}^* \end{cases}$$

THE SOLUTIONS ARE:

$$\Gamma_{ML} = \frac{\frac{B_2}{C_2} \pm \sqrt{\left(\frac{B_2}{C_2}\right)^2 - 4\frac{C_2^*}{C_2}}}{2} = \frac{B_2 \pm \sqrt{B_2 - 4|C_2|^2}}{2C_2}$$

SIMILARLY, SOLVING (1) FOR Γ_{ML} AND EQUATING THE RESULT TO (2) GIVES: $\Gamma_{MS} = \dfrac{B_1 \pm \sqrt{B_1 - 4|C_1|^2}}{2C_1}$ WHERE $\begin{cases} B_1 = 1 + |S_{11}|^2 - |S_{22}|^2 - |\Delta|^2 \\ C_1 = S_{11} - \Delta S_{22}^* \end{cases}$

(b) FOR $S_{12} \to 0$ IT FOLLOWS FROM (3.6.3) AND (3.6.4) THAT

$$\Gamma_{MS} = \left(S_{11} + \frac{S_{12}S_{21}\Gamma_{ML}}{1 - S_{22}\Gamma_{ML}}\right)^* \approx S_{11}^*$$

AND

$$\Gamma_{ML} = \left(S_{22} + \frac{S_{12}S_{21}\Gamma_{MS}}{1 - S_{11}\Gamma_{MS}}\right)^* \approx S_{22}^*$$

3.20) (a) $|C_1|^2 = |S_{11} - \Delta S_{22}^*|^2 = (S_{11} - \Delta S_{22}^*)(S_{11}^* - \Delta^* S_{22})$

$= |S_{11}|^2 - S_{11}S_{22}\Delta^* - S_{11}^*S_{22}^*\Delta + |\Delta S_{22}|^2$

$= |S_{11}|^2 - |S_{11}S_{22}|^2 + S_{11}S_{22}S_{12}^*S_{21}^* - |S_{11}S_{22}|^2 + S_{11}^*S_{22}^*S_{12}S_{21} + |\Delta S_{22}|^2$ (1)

SINCE $|\Delta|^2 = (S_{11}S_{22} - S_{12}S_{21})(S_{11}^*S_{22}^* - S_{12}^*S_{21}^*)$

$= |S_{11}S_{22}|^2 - S_{11}S_{22}S_{12}^*S_{21}^* - S_{11}^*S_{22}^*S_{12}S_{21} + |S_{12}S_{21}|^2$

THEN: $|S_{11}S_{22}|^2 - S_{11}S_{22}S_{12}^*S_{21}^* - S_{11}^*S_{22}^*S_{12}S_{21} = |\Delta|^2 - |S_{12}S_{21}|^2$ (2)

(2) INTO (1): $|C_1|^2 = |S_{11}|^2 - |\Delta|^2 + |S_{12}S_{21}|^2 - |S_{11}S_{22}|^2 + |\Delta|^2|S_{22}|^2$

$= |S_{12}S_{21}|^2 + |S_{11}|^2(1 - |S_{22}|^2) - |\Delta|^2(1 - |S_{22}|^2)$

$= |S_{12}S_{21}|^2 + (1 - |S_{22}|^2)(|S_{11}|^2 - |\Delta|^2)$

SIMILARLY: $|C_2|^2 = |S_{12}S_{21}|^2 + (1 - |S_{11}|^2)(|S_{22}|^2 - |\Delta|^2)$

3.21) $K = 1.03$ AND $\Delta = 0.324\angle{-64.8^\circ}$ $\therefore$ UNCONDITIONALLY STABLE

FROM (3.6.5) : $\Gamma_{Ms} = 0.73\angle{135^\circ}$

FROM (3.6.6) : $\Gamma_{ML} = 0.95\angle{33.8^\circ}$

FROM (3.6.10) : $G_{T,max} = 19.08$ OR 12.8 dB

A DESIGN USING LUMPED ELEMENTS IS :

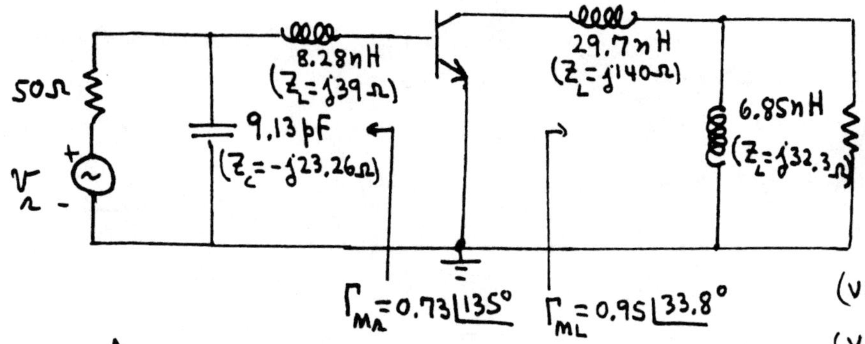

$f = 750$ MHz

$(VSWR)_{in} = 1$

$(VSWR)_{out} = 1$

A DESIGN USING MICROSTRIP LINES IS :

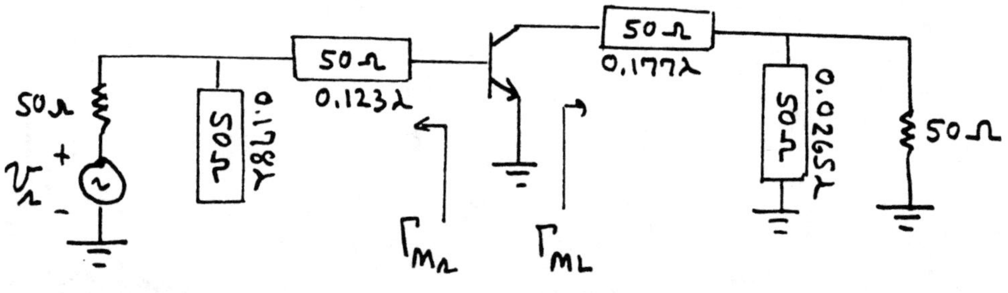

3.22) (a) THE ADMITTANCE OF THE 70 Ω STUB IS : $y_{oc} = j$ OR $Y_{oc} = \frac{j}{70}$

THE ADMITTANCE OF THE TWO PARALLEL STUBS IS :

$$Y_{oc,TOTAL} = \frac{j}{70} + \frac{j}{70} = \frac{2j}{70} = \frac{j}{35} = j29 \text{ mS}$$

THE ADMITTANCE OF THE 50-Ω LOAD PLUS $Y_{oc,TOTAL}$ IS (CALLED Y_X):

$$Y_X = \frac{1}{50} + \frac{j}{35} = 20 + j29 \text{ mS}$$

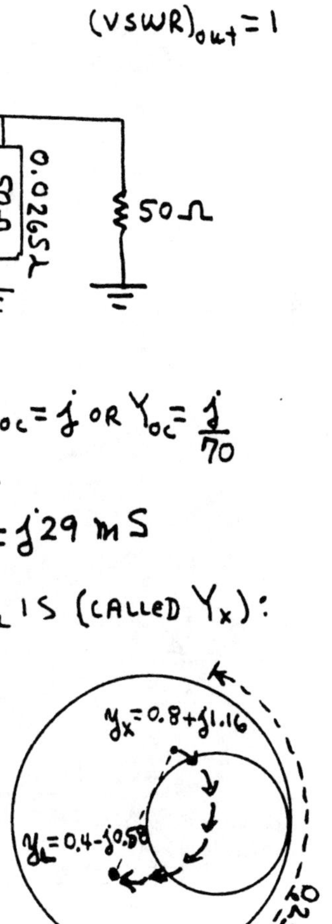

NORMALIZING Y_X WITH $Y_{o1} = \frac{1}{Z_{01}} = \frac{1}{40}$ $(y_X = 0.8 + j1.16)$

AND ROTATING IN THE SMITH CHART

A DISTANCE $l = 0.25\lambda$ GIVES

$z_L = \frac{1}{y_L} = 0.8 + j1.16$ $\therefore \Gamma_L = \Gamma_{ML} = 0.55\angle{67^\circ}$

(b) FROM FIG. 2.5.4 (WITH $\varepsilon_r = 10$ AND $h = 30$ mils):

$W = 28.5$ mils AND $\varepsilon_{ff} = 6.68$, $\lambda_o = \frac{3\,10^{10}}{2\,10^9} = 15$ cm.

$\therefore l_1 = 0.25\lambda = \frac{0.25\lambda_o}{\sqrt{\varepsilon_{ff}}} = \frac{0.25(15)}{\sqrt{6.68}} = 1.45$ cm.

3.23) THE OUTPUT IS MATCHED WITH $\Gamma_{ML} = \Gamma_{OUT}^* = 0.718 \angle 103.9°$. THEREFORE, $\Gamma_x = 0$ (OR $Z_x = 50\,\Omega$) AND $(VSWR)_{out} = 1$.

NOTE: THIS IS THE OUTPUT MATCHING NETWORK IN FIG. 3.6.4 (SEE FIG. 3.6.3b). IT IS SIMPLE TO VERIFY THAT $\Gamma_x = 0$ USING FIG. 3.6.3b. THAT IS, PLOT $y_{out} = y_{ML}^* = 0.414 + j1.19$ IN THE SMITH CHART AND FIND y_x, WHICH WILL BE $y_x = 1$ ($Z_x = 50\,\Omega$ OR $\Gamma_x = 0$).

3.24) (a) SINCE $K = 1.19$ AND $\Delta = 0.399 \angle 126.5°$ (i.e., $|\Delta| < 1$) THE BJT IS UNCONDITIONALLY STABLE AT 3.5 GHz. THEREFORE, IT CAN BE DESIGNED FOR A SIMULTANEOUS CONJUGATE MATCH.

(b) FROM (3.6.5) AND (3.6.6): $\Gamma_{Ms} = 0.798 \angle -166.9°$, $\Gamma_{ML} = 0.904 \angle 68°$

FROM (3.6.10): $G_{T,max} = 23.06$ OR 13.63 dB.

A DESIGN FOR THE AMPLIFIER AT 3.5 GHz IS:

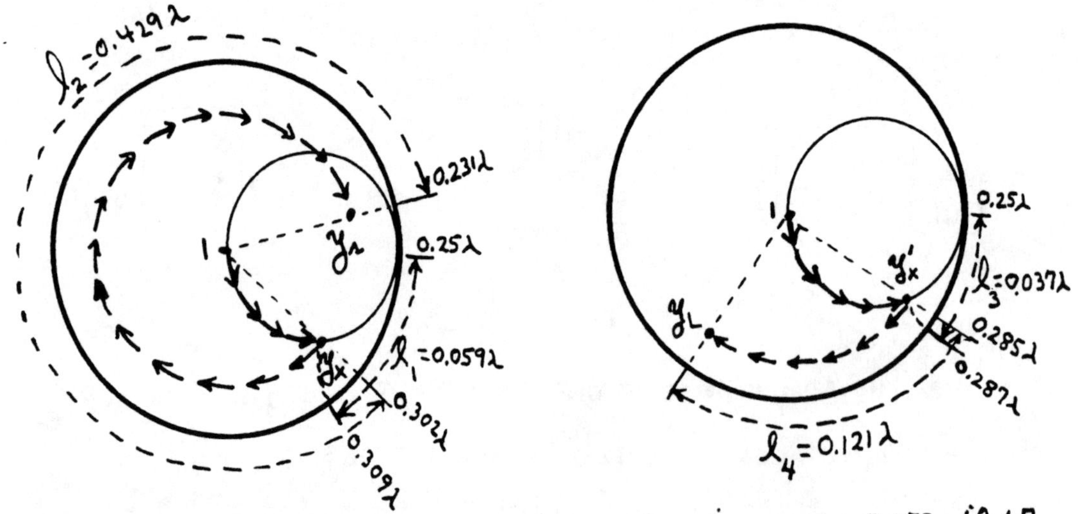

$y_x = 1 - j2.6$, $y_s = 4.4 + j4.4$ $\qquad$ $y_x' = 1 - j4.2$, $y_L = 0.073 - j0.67$

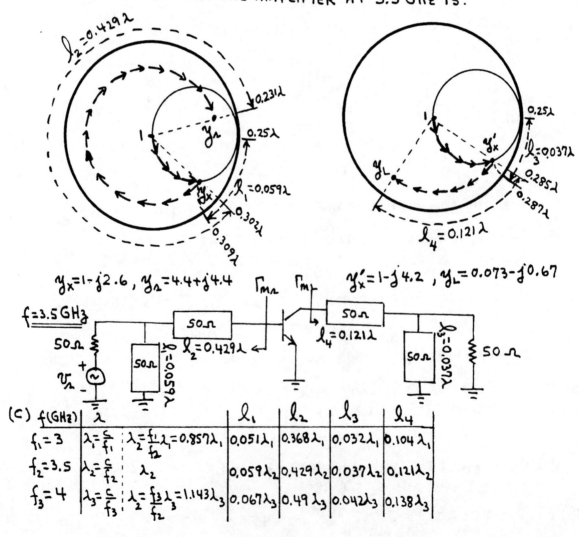

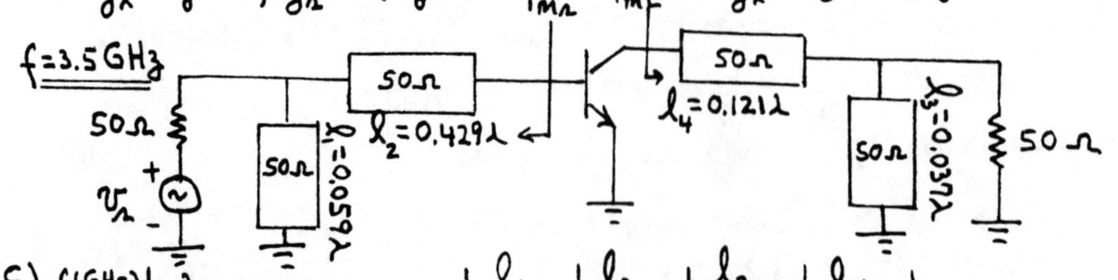

(c)

f(GHz)	λ		l_1	l_2	l_3	l_4
$f_1 = 3$	$\lambda_1 = \frac{c}{f_1}$	$\lambda_2 = \frac{f_1}{f_2}\lambda_1 = 0.857\lambda_1$	$0.051\lambda_1$	$0.368\lambda_1$	$0.032\lambda_1$	$0.104\lambda_1$
$f_2 = 3.5$	$\lambda_2 = \frac{c}{f_2}$	λ_2	$0.059\lambda_2$	$0.429\lambda_2$	$0.037\lambda_2$	$0.121\lambda_2$
$f_3 = 4$	$\lambda_3 = \frac{c}{f_3}$	$\lambda_2 = \frac{f_3}{f_2}\lambda_3 = 1.143\lambda_3$	$0.067\lambda_3$	$0.49\lambda_3$	$0.042\lambda_3$	$0.138\lambda_3$

USING THE SMITH CHART IT IS SIMPLE TO FIND THE VALUES OF Γ_s AND Γ_L AT f_1 AND f_3. THE VALUES OF G_T ARE CALCULATED USING (3.2.1). THE RESULTS ARE:

f(GHz)	Γ_s	Γ_L	G_T
3	$0.833\angle{-118.5°}$	$0.934\angle{82.2°}$	0.944 OR −0.25 dB
3.5	$0.798\angle{-166.9°}$	$0.904\angle{68°}$	23.06 OR 13.63 dB
4	$0.749\angle{145.6°}$	$0.878\angle{51.5°}$	2.257 OR 3.54 dB

3.25) (a) THE TRANSISTOR IS UNCONDITIONALLY STABLE ($K = 1.033$, $\Delta = 0.324\angle{-64.8°}$)

WITH $g_p = \dfrac{G_p}{|S_{21}|^2} = \dfrac{10^1}{(1.92)^2} = 2.713$, WE OBTAIN FROM (3.7.4) AND (3.7.5):

$$C_p = 0.781\angle{33.85°} \quad \text{AND} \quad r_p = 0.214$$

THE $G_p = 10\,dB$ GAIN CIRCLE IS DRAWN IN THE SMITH CHART. SELECTING Γ_L AT POINT "A": $\Gamma_L = 0.567\angle{33.85°}$, GIVES

$$\Gamma_s = \Gamma_{IN}^* = 0.276\angle{93.33°}$$

AND

$$\Gamma_{OUT} = 0.86\angle{-33.85°}$$

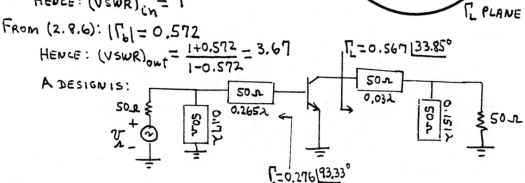

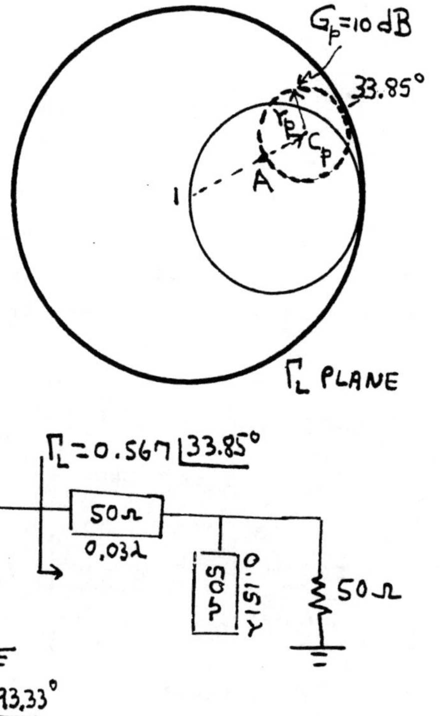

FROM (2.8.3): $|\Gamma_a| = 0$ (SINCE $\Gamma_s = \Gamma_{IN}^*$)

HENCE: $(VSWR)_{in} = 1$

FROM (2.8.6): $|\Gamma_b| = 0.572$

HENCE: $(VSWR)_{out} = \dfrac{1+0.572}{1-0.572} = 3.67$

A DESIGN IS:

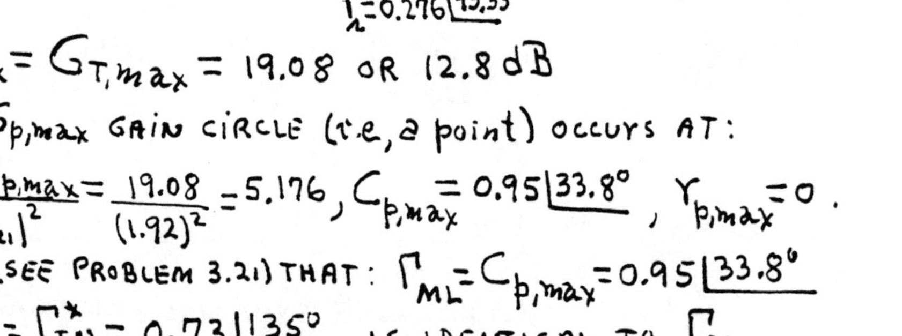

(b) $G_{p,max} = G_{T,max} = 19.08$ OR 12.8 dB

THE $G_{p,max}$ GAIN CIRCLE (i.e, a point) OCCURS AT:

$$g_{p,max} = \frac{G_{p,max}}{|S_{21}|^2} = \frac{19.08}{(1.92)^2} = 5.176, \quad C_{p,max} = 0.95\angle{33.8°}, \quad r_{p,max} = 0.$$

OBSERVE (SEE PROBLEM 3.21) THAT: $\Gamma_{ML} = C_{p,max} = 0.95\angle{33.8°}$

AND $\Gamma_s = \Gamma_{IN}^* = 0.73\angle{135°}$ IS IDENTICAL TO Γ_{Ms}.

3.26) FOR THIS TRANSISTOR: $K = 1.053$ AND $\Delta = 0.576\angle{-85.4^\circ}$. THEREFORE, IT IS UNCONDICIONALLY STABLE.

$$G_{p,max} = G_{T,MAX} = 77.12 \text{ OR } 18.87 \text{ dB}$$

$G_p = 10$ dB CIRCLE: $g_p = 0.977$, $C_p = 0.306\angle{39.45^\circ}$, $r_p = 0.693$

THE $G_p = 10$ dB CONSTANT-GAIN CIRCLE IS SHOWNIN THE SMITH CHART. THE Γ_L SELECTED IS SHOWN AS "A": $\Gamma_L = 0.85\angle{89^\circ}$.

THEN: $\Gamma_s = \Gamma_{IN}^* = 0.793\angle{64.2^\circ}$, $\Gamma_{OUT} = 0.798\angle{-42.3^\circ}$

$|\Gamma_a| = 0$, $(VSWR)_{in} = 1$

$|\Gamma_b| = 0.9$, $(VSWR)_{out} = 18.9$

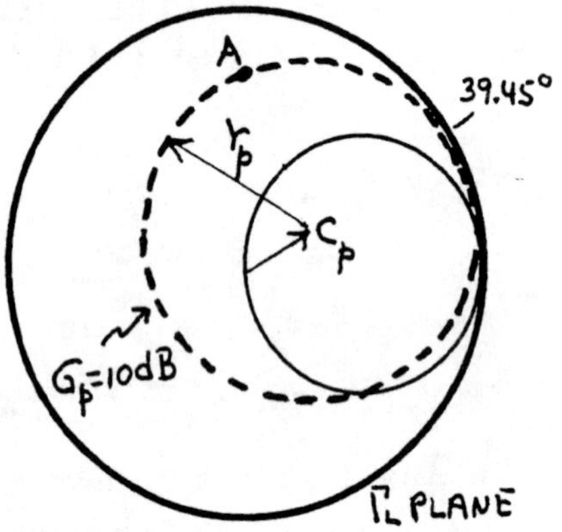

3.27) (a) $K = 0.53$, $\Delta = 0.524\angle{-142.9^\circ}$; IT IS POTENTIALLY UNSTABLE.

OUTPUT STABILITY CIRCLE [(3.3.7) AND (3.3.8)]:

$$C_L = 1.47\angle{76.6^\circ} \text{ AND } r_L = 0.668$$

$G_p = 10$ dB CONSTANT-GAIN CIRCLE [(3.7.4) AND (3.7.5)]: $g_p = \frac{10^1}{(2.43)^2} = 1.694$,

$$C_p = 0.41\angle{76.6^\circ} \text{ AND } r_p = 0.641$$

(b) THE $G_p = 10$ dB GAIN CIRCLE IS DRAWN ON THE SMITH CHART. THREE VALUES OF Γ_L ARE DENOTED BY "a", "b", AND "c". THEN,

	Γ_L	$\Gamma_s = \Gamma_{IN}^*$	Γ_{OUT}	$\|\Gamma_b\|$	$(VSWR)_{in}$	$(VSWR)_{out}$
"a"	$0.59\angle{0^\circ}$	$0.687\angle{106.11^\circ}$	$0.84\angle{-70.73^\circ}$	0.889	1	17
"b"	$0.24\angle{-90^\circ}$	$0.757\angle{100.35^\circ}$	$0.83\angle{-76.22^\circ}$	0.891	1	17.3
"c"	$0.67\angle{145^\circ}$	$0.851\angle{97.7^\circ}$	$0.84\angle{-83.74^\circ}$	0.889	1	17

THE 3 VALUES OF Γ_s ARE IN THE STABLE REGION, SINCE $C_s = 1.3\angle{115.7^\circ}$ AND $r_s = 0.46$.

(c) FOR $G_p = 15$ dB: $g_p = \frac{10^{1.5}}{(2.43)^2} = 5.355$, $C_p = 0.81\angle{76.6^\circ}$, $r_p = 0.402$

FOR $G_p = 20$dB: $g_p = \frac{10^2}{(2.43)^2} = 16.94$, $C_p = 1.17\angle{76.6^\circ}$, $r_p = 0.457$

FOR $G_p = 40$ dB: $g_p = \frac{10^4}{(2.43)^2} = 1{,}694$, $C_p = 1.46\angle{76.6^\circ}$, $r_p = 0.665$

THE $G_p = 15$ dB, 20 dB, AND 40 dB GAIN CIRCLES ARE ALSO DRAWN ON THE SMITH CHART. THE $G_p = 40$ dB CIRCLE ALMOST COINCIDES WITH THE OUTPUT STABILITY CIRCLE.

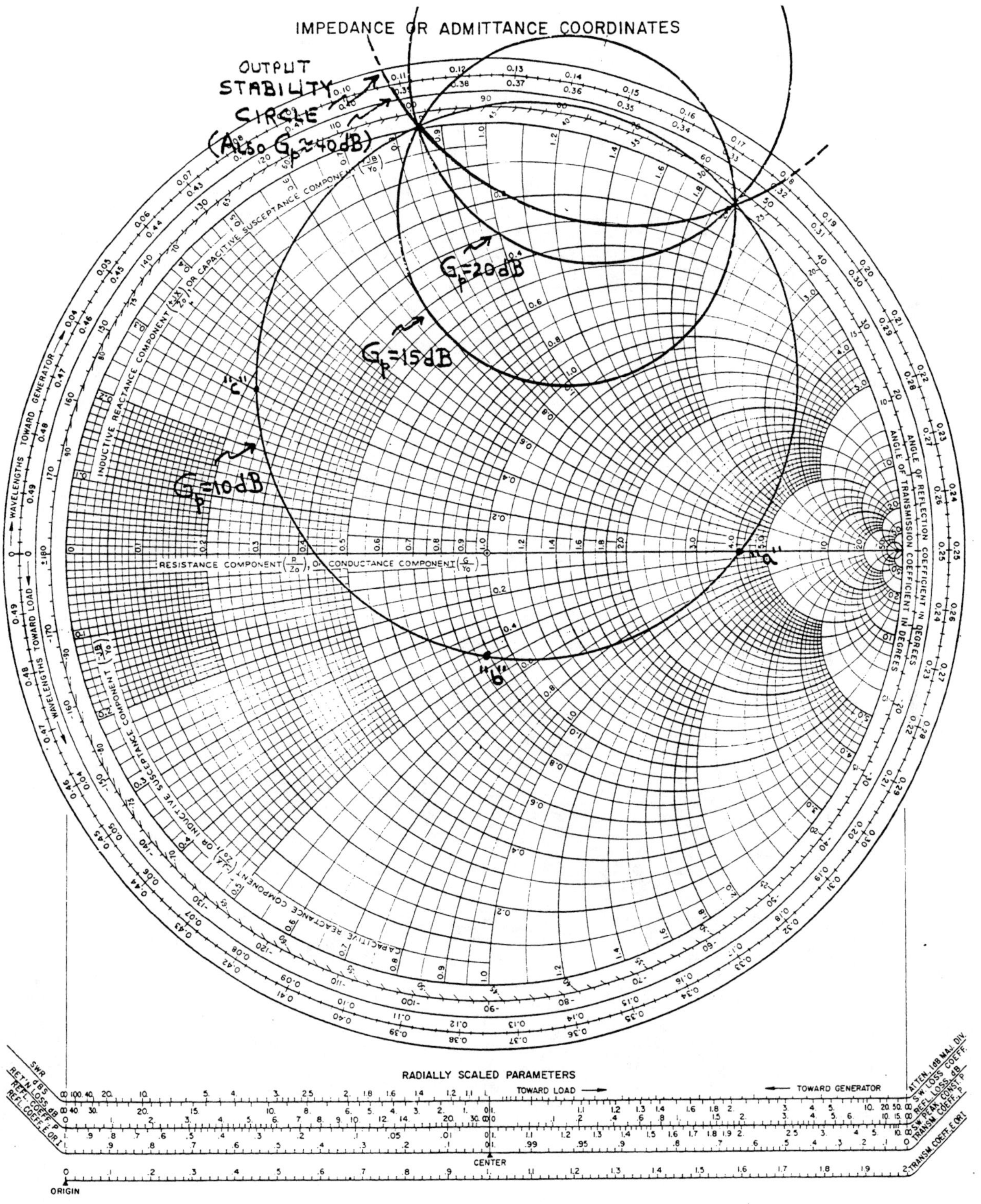
IMPEDANCE OR ADMITTANCE COORDINATES
OUTPUT STABILITY CIRCLE
(Also G_p = 40dB)
G_p = 20dB
G_p = 15dB
G_p = 10dB
"c"
"a"
"b"
RESISTANCE COMPONENT (R/Zo), OR CONDUCTANCE COMPONENT (G/Yo)
RADIALLY SCALED PARAMETERS
TOWARD LOAD
TOWARD GENERATOR
CENTER
ORIGIN

3.28) OUTPUT STABILITY CIRCLE: $|\Gamma_L - C_L|^2 = r_L^2$

OR
$$|\Gamma_L|^2 - \Gamma_L C_L^* - \Gamma_L^* C_L + |C_L|^2 = r_L^2 \qquad (1)$$

(1) INTERSECTS THE SMITH CHART WHEN $|\Gamma_L| = 1$; THEN

$$1 - \Gamma_L C_L^* - \Gamma_L^* C_L + |C_L|^2 = r_L^2$$

LET $\Gamma_L = U_L + jV_L$ AND $C_L = \mathrm{Re}[C_L] + j\,\mathrm{Im}[C_L]$:

$$-(U_L + jV_L)(\mathrm{Re}[C_L] - j\,\mathrm{Im}[C_L]) - (U_L - jV_L)(\mathrm{Re}[C_L] + j\,\mathrm{Im}[C_L]) = r_L^2 - |C_L|^2 - 1$$

$$2U_L \mathrm{Re}[C_L] + 2V_L \mathrm{Im}[C_L] = 1 + |C_L|^2 - r_L^2$$

OR

$$V_L = -\frac{\mathrm{Re}[C_L]}{\mathrm{Im}[C_L]} U_L + \frac{1 + |C_L|^2 - r_L^2}{2\,\mathrm{Im}[C_L]} \qquad (2)$$

(2) IS THE EQUATION OF A STRAIGHT LINE WITH SLOPE $m_L = -\dfrac{\mathrm{Re}[C_L]}{\mathrm{Im}[C_L]}$ AND INTERCEPT $b_L = \dfrac{1 + |C_L|^2 - r_L^2}{2\,\mathrm{Im}[C_L]}$. THIS LINE INTERSECTS THE SMITH CHART (i.e., $|\Gamma_L| = 1$) AT TWO POINTS.

SIMILARLY, THE POWER GAIN CIRCLE INTERSECTS $|\Gamma_L| = 1$ AT TWO POINTS DETERMINED BY: (LET $C_p = \mathrm{Re}[C_p] + j\,\mathrm{Im}[C_p]$)

$$V_L = -\frac{\mathrm{Re}[C_p]}{\mathrm{Im}[C_p]} U_L + \frac{1 + |C_p|^2 - r_p^2}{2\,\mathrm{Im}[C_p]} = -m_p U_L + b_p$$

THE POINTS OF INTERSECTION DETERMINED BY (2) AND (3) ARE EQUAL IF $m_L = m_p$ AND $b_L = b_p$.

$$m_L = -\frac{\mathrm{Re}[C_L]}{\mathrm{Im}[C_L]} = -\frac{\mathrm{Re}[S_{22}^* - \Delta^* S_{11}]}{\mathrm{Im}[S_{22}^* - \Delta^* S_{11}]}$$

$$m_p = -\frac{\mathrm{Re}[C_p]}{\mathrm{Im}[C_p]} = -\frac{\mathrm{Re}[C_2^*]}{\mathrm{Im}[C_2^*]} = -\frac{\mathrm{Re}[S_{22}^* - \Delta^* S_{11}]}{\mathrm{Im}[S_{22}^* - \Delta^* S_{11}]}$$

$$\therefore\ m_L = m_p$$

IT ALSO FOLLOWS THAT $b_L = b_p$.

3.29) (a) $K = 1.032$, $\Delta = 1.65\angle 78.1^\circ$ ∴ POTENTIALLY UNSTABLE

FROM (3.3.7) TO (3.3.10):

INPUT STAB. CIRCLE $\begin{cases} C_s = 0.285\angle -161.1^\circ \\ r_s = 0.65 \end{cases}$ OUTPUT STAB. CIRCLE $\begin{cases} C_L = 0.315\angle -61.5^\circ \\ r_L = 0.627 \end{cases}$

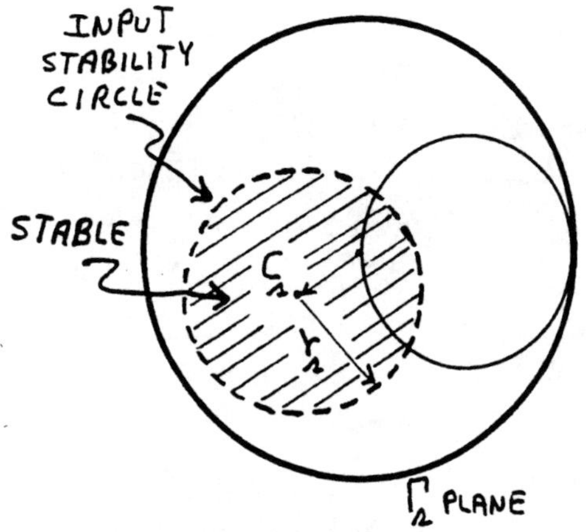

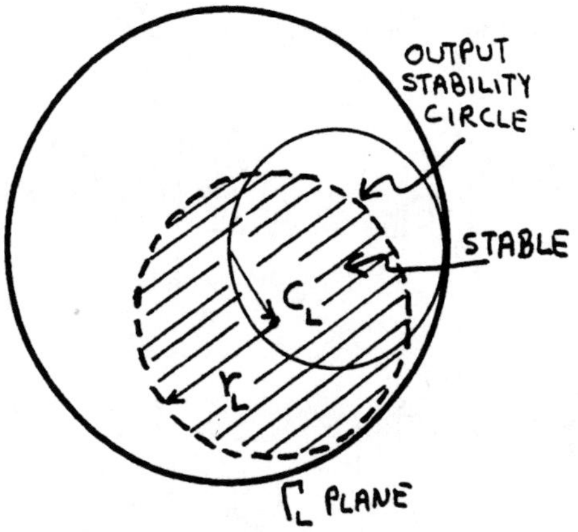

(b) FROM (3.2.3) WITH $\Gamma_L = 0$: (AND $\Gamma_{IN} = S_{11}$ WHEN $\Gamma_L = 0$)

$$G_p = \frac{|S_{21}|^2}{1-|S_{11}|^2} = \frac{4^2}{1-(0.5)^2} = 21.33 \text{ OR } 13.3\,\text{dB}$$

(c) G_p CAN BE INFINITE, BECAUSE IT IS POTENTIALLY UNSTABLE. AS Γ_L APPROACHES THE STABILITY CIRCLE, $G_p \to \infty$.

3.30) FROM EXAMPLE 3.8.1: $G_{A,max} = 9.66\,\text{dB}$, $G_A = 9.66 - 2 = 7.66\,\text{dB}$.

FROM (3.7.15) AND (3.7.16), FOR THE $G_A = 7.66\,\text{dB}$ GAIN CIRCLE:

$$g_a = \frac{10^{0.766}}{(2.3)^2} = 1.103, \quad C_a = 0.503\angle -40.45^\circ, \quad r_a = 0.436$$

THE VALUES OF Γ_s ON THE 7.6 dB CIRCLE ARE GIVEN BY:

$$\Gamma_s = C_a + r_a e^{j\theta_1} = 0.503\angle -40.45^\circ + 0.436 e^{j\theta_1}$$

LETTING $\theta_1 = 0, \frac{\pi}{2}, \pi$, AND $\frac{3\pi}{2}$ WE OBTAIN THE VALUES OF Γ_s SHOWN IN THE TABLE. THE ASSOCIATED VALUES OF Γ_{OUT} ARE ALSO SHOWN.

FOR $(VSWR)_{out} = 1.5$ (OR $|\Gamma_b| = 0.2$), USING (3.8.7) AND (3.8.8), THE CENTER AND RADIUS OF THE $(VSWR)_{out} = 1.5$ CIRCLE ARE CALCULATED.

THE VALUES OF Γ_L ON THE $(VSWR)_{out} = 1.5$ CIRCLE ARE GIVEN BY:

$\Gamma_L = C_{V_o} + r_{V_o} e^{j\theta_2}$. LETTING $\theta_2 = 0, \frac{\pi}{2}, \pi$, AND $\frac{3\pi}{2}$, FOUR VALUES OF Γ_L ON THE $(VSWR)_{out} = 1.5$ CIRCLE ARE CALCULATED, AS SHOWN IN THE TABLE. THE ASSOCIATED VALUES OF Γ_{IN}, $|\Gamma_a|$, AND $(VSWR)_{in}$ ARE ALSO SHOWN.

Γ_s	Γ_{OUT}	$(VSWR)_{out}=1.5$ CIRCLE	Γ_L	Γ_{IN}	$\lvert\Gamma_a\rvert$	$(VSWR)_{in}$
$(\theta_1=0^\circ)$ $0.881\angle -21.73^\circ$	$0.451\angle 176.74^\circ$	$C_{V_o}=0.437\angle -176.74^\circ$ $r_{V_o}=0.161$	$0.276\angle -174.8^\circ$ $(\theta_2=0)$	$0.683\angle 33.8$	0.596	3.95
			$0.457\angle 162.7^\circ$ $(\theta_2=\frac{\pi}{2})$	$0.712\angle 28.8^\circ$	0.507	3.05
			$0.598\angle -177.6^\circ$ $(\theta_2=\pi)$	$0.771\angle 30.9^\circ$	0.495	2.96
			$0.474\angle -156.9^\circ$ $(\theta_2=\frac{3\pi}{2})$	$0.748\angle 35.7^\circ$	0.604	4.05
$(\theta_1=\pi/2)$ $0.398\angle 15.99^\circ$	$0.005\angle 95.36^\circ$	$C_{V_o}=0.005\angle -95.36^\circ$ $r_{V_o}=0.2$	$0.2\angle -1.4^\circ$ $(\theta_2=0)$	$0.539\angle 38.5^\circ$	0.501	3.00
			$0.195\angle 90.1^\circ$ $(\theta_2=\frac{\pi}{2})$	$0.582\angle 30.3^\circ$	0.491	2.93
			$0.2\angle -178.6^\circ$ $(\theta_2=\pi)$	$0.659\angle 34^\circ$	0.591	3.89
			$0.2\angle -90.1^\circ$ $(\theta_2=\frac{3\pi}{2})$	$0.628\angle 41.5^\circ$	0.598	3.98
$(\theta_1=\pi)$ $0.331\angle -99.26^\circ$	$0.249\angle 62.12^\circ$	$C_{V_o}=0.24\angle -62.12^\circ$ $r_{V_o}=0.188$	$0.368\angle -35.2^\circ$ $(\theta_2=0)$	$0.539\angle 47^\circ$	0.473	2.79
			$0.115\angle -12.1^\circ$ $(\theta_2=\frac{\pi}{2})$	$0.569\angle 38^\circ$	0.543	3.38
			$0.225\angle -109.6^\circ$ $(\theta_2=\pi)$	$0.652\angle 40.8^\circ$	0.613	4.17
			$0.416\angle -74.3^\circ$ $(\theta_2=\frac{3\pi}{2})$	$0.633\angle 48.8^\circ$	0.56	3.54
$(\theta_1=3\pi/2)$ $0.853\angle -63.34^\circ$	$0.547\angle 94.15^\circ$	$C_{V_o}=0.531\angle -94.15$ $r_{V_o}=0.142$	$0.54\angle -78.9^\circ$ $(\theta_2=0)$	$0.666\angle 51.8^\circ$	0.526	3.22
			$0.39\angle -95.6^\circ$ $(\theta_2=\frac{\pi}{2})$	$0.673\angle 45.7^\circ$	0.606	4.07
			$0.559\angle -108.8^\circ$ $(\theta_2=\pi)$	$0.742\angle 46.3^\circ$	0.597	3.97
			$0.673\angle -93.3^\circ$ $(\theta_2=\frac{3\pi}{2})$	$0.74\angle 51.9^\circ$	0.485	2.88

FROM THESE CALCULATIONS IT IS SEEN THAT A DESIGN WITH $\Gamma_s = 0.331\angle -99.26^\circ$ AND $\Gamma_L = 0.368\angle -35.2^\circ$ RESULTS IN $(VSWR)_{in}=2.79$ AND $(VSWR)_{out}=1.5$.

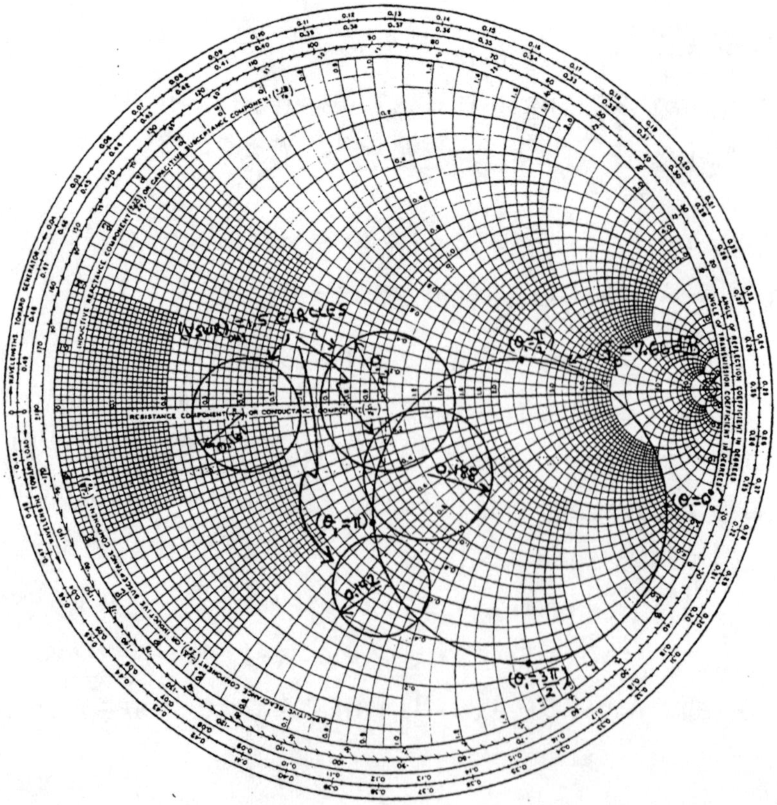

3.31) (a) $K=1.344$, $\Delta=2.156\angle -16.6°$ ∴ POTENTIALLY UNSTABLE

(b) OUTPUT STABILITY CIRCLE: $C_L=0.261\angle -36.3°$, $r_L=0.409$

(c) FROM (3.6.6) (USING THE + SIGN): $\Gamma_{ML}=0.319\angle -36.3°$

(d) $G_{p,min}=G_{T,min}=44.82$ OR 16.51 dB

(e)

G_p	CENTER AND RADIUS
18 dB	$C_p=0.3\angle -36.3°$, $r_p=0.234$
21 dB	$C_p=0.28\angle -36.3°$, $r_p=0.34$
26 dB	$C_p=0.266\angle -36.3°$, $r_p=0.39$
36 dB	$C_p=0.261\angle -36.3°$, $r_p=0.407$

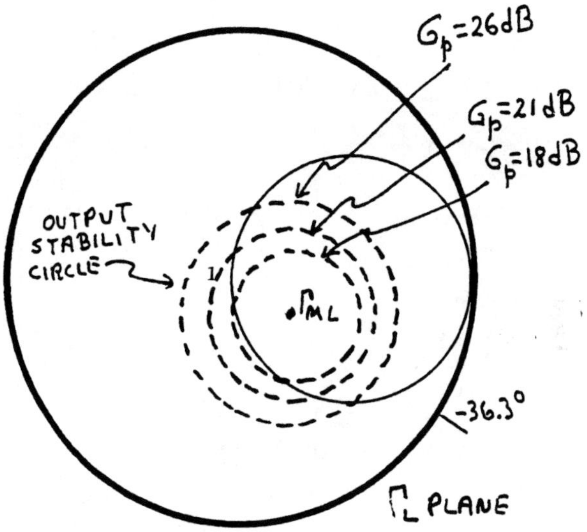

THE 26 dB GAIN CIRCLE, THE 36 dB GAIN CIRCLE, AND THE OUTPUT STABILITY CIRCLE ALMOST COINCIDE.

(f) INPUT STABILITY CIRCLE: $C_s=0.102\angle 107.4°$, $r_s=0.44$

FROM (3.6.5) (USING THE + SIGN): $\Gamma_{Ms}=0.127\angle 107.44°$

(g) $(VSWR)_{in}=(VSWR)_{out}=1$.

3.32) (a) $K=1.17$, $\Delta=0.368\angle 27.91°$ UNCONDITIONALLY STABLE

$G_{p,max}=9.24$ OR 9.66 dB, $G_p=9.66-1=8.66$ dB

FOR THE $G_p=8.66$ dB CIRCLE: $g_p=1.388$, $C_p=0.292\angle -129.4°$, $r_p=0.466$

(b) THE VALUES OF Γ_L ON THE 8.66 dB CIRCLE ARE:

$\Gamma_L=C_p+r_p e^{j\theta_1}=0.292\angle -129.4°+0.466 e^{j\theta_1}$

LETTING $\theta_1=0, \frac{\pi}{2}, \pi$, AND $\frac{3\pi}{2}$ WE OBTAIN THE VALUES OF Γ_L SHOWN IN THE TABLE. THE ASSOCIATED VALUES OF Γ_{IN} ARE ALSO SHOWN.

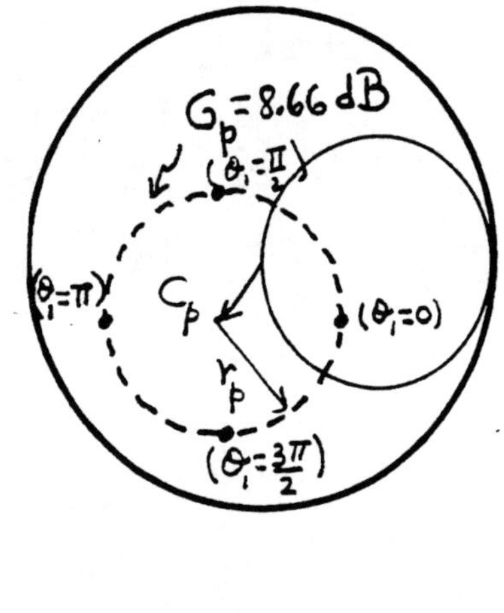

FOR $(VSWR)_{in}=1.5$ (OR $|\Gamma_a|=0.2$), USING (3.8.3) AND (3.8.4), THE CENTER AND RADIUS OF THE $(VSWR)_{in}=1.5$ CIRCLE ARE CALCULATED.

(c) THE VALUES OF Γ_s ON THE $(VSWR)_{in} = 1.5$ CIRCLE ARE GIVEN BY $\Gamma_s = C_{V_i} + r_{V_i} e^{j\theta_2}$. LETTING $\theta_2 = 0$ AND π, TWO VALUES OF Γ_s ON THE $(VSWR)_{in} = 1.5$ CIRCLE ARE CALCULATED (SEE THE TABLE). THE ASSOCIATED VALUES OF Γ_{OUT}, $|\Gamma_b|$, AND $(VSWR)_{OUT}$ ARE ALSO SHOWN

Γ_L	Γ_{IN}	$(VSWR)_{in} = 1.5$ CIRCLE	Γ_s	Γ_{OUT}	$\|\Gamma_b\|$	$(VSWR)_{out}$
$(\theta_1 = 0)$ 0.36 $\angle -38.8°$	0.548 $\angle 47.1°$	$C_{V_i} = 0.532 \angle -47.1°$	0.637 $\angle -37.7°$ $(\theta_2 = 0)$	0.311 $\angle 129.5°$	0.475	2.8
		$r_{V_i} = 0.142$	0.448 $\angle -60.5°$ $(\theta_2 = \pi)$	0.259 $\angle 91.2°$	0.304	1.875
$(\theta_1 = \pi/2)$ 0.304 $\angle 127.6°$	0.634 $\angle 27.9°$	$C_{V_i} = 0.619 \angle -27.9°$	0.729 $\angle -23.4°$ $(\theta_2 = 0)$	0.314 $\angle 161.7°$	0.368	2.16
		$r_{V_i} = 0.122$	0.514 $\angle -34.3°$ $(\theta_2 = \pi)$	0.219 $\angle 123.5°$	0.419	2.44
$(\theta_1 = \pi)$ 0.689 $\angle -160.9°$	0.81 $\angle 34.2°$	$C_{V_i} = 0.799 \angle -34.2°$	0.859 $\angle -31.5°$ $(\theta_2 = 0)$	0.501 $\angle 153.7°$	0.306	1.88
		$r_{V_i} = 0.071$	0.741 $\angle -37.3°$ $(\theta_2 = \pi)$	0.399 $\angle 136.2°$	0.483	2.87
$(\theta_1 = 3\pi/2)$ 0.716 $\angle -105°$	0.782 $\angle 49.4°$	$C_{V_i} = 0.77 \angle -49.4°$	0.824 $\angle -45.2°$ $(\theta_2 = 0)$	0.518 $\angle 124.3°$	0.430	2.51
		$r_{V_i} = 0.08$	0.721 $\angle -54.2°$ $(\theta_2 = \pi)$	0.429 $\angle 107.5°$	0.415	2.42

FROM THESE CALCULATIONS IT IS SEEN THAT A DESIGN WITH $\Gamma_L = 0.689 \angle -160.9°$ AND $\Gamma_s = 0.859 \angle -31.5°$ GIVES: $G_p = 8.66$ dB, $(VSWR)_{in} = 1.88$, $(VSWR)_{out} = 1.5$.

3.33) (a) FROM EXAMPLE 3.7.2, FOR THE $G_p = 10$ dB CIRCLE: $C_p = 0.572 \angle 97.2°$, $r_p = 0.473$

FROM (3.8.9) AND (3.8.10), WITH $C_{oo} = C_p$ AND $r_{oo} = r_p$, WE OBTAIN:

$$C_i = 1.131 \angle 170.6° \text{ AND } r_i = 0.622$$

(b) FOR $\Gamma_L = 0.1 \angle 97°$, $\Gamma_{IN} = 0.52 \angle 179.32°$,

$\Gamma_s = \Gamma_{IN}^* = 0.52 \angle -179.32°$

THEN $(VSWR)_{in} = 1$

FOR $(VSWR)_{in} = 2$ (OR $|\Gamma_a| = 0.333$), WE OBTAIN FROM (3.8.3) AND (3.8.4):

$C_{V_i} = 0.477 \angle 179.32°$, $r_{V_i} = 0.251$

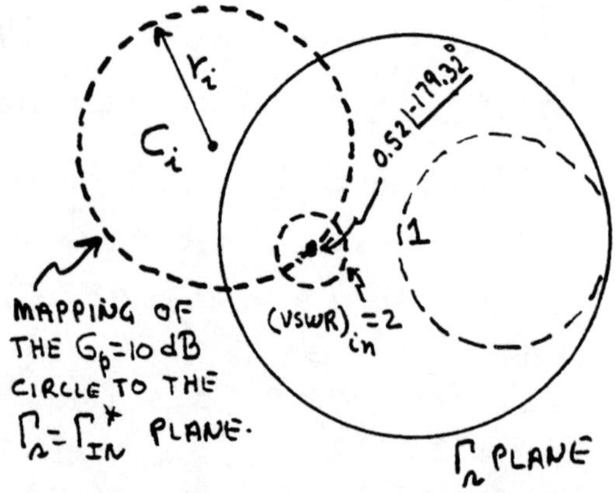

3.34) $K = 0.924$, $\Delta = 0.137\angle{-139^\circ}$ $\therefore$ POTENTIALLY UNSTABLE

$$G_{MSG} = \frac{|S_{21}|}{|S_{12}|} = \frac{8}{0.03} = 266.6 \text{ OR } 24.3\,dB$$

DESIGN FOR $G_p = 20\,dB$ (i.e., 4.3 dB LESS THAN G_{MSG}).

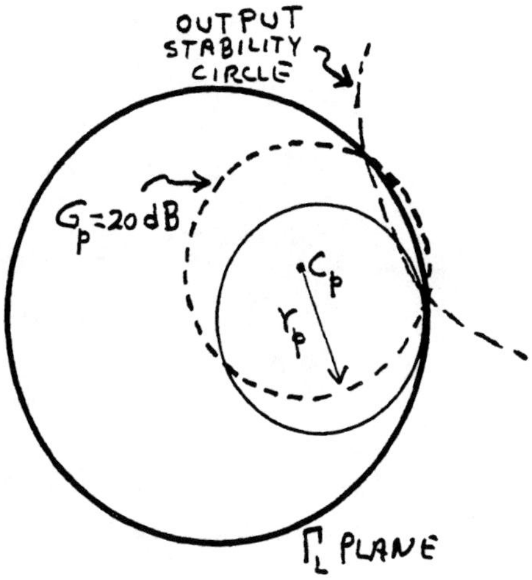

OUTPUT STABILITY CIRCLE:

$C_L = 2.26\angle{40.1^\circ}$, $r_L = 1.307$

$G_p = 20\,dB$ CONSTANT-GAIN CIRCLE: ($g_p = 1.563$)

$C_p = 0.505\angle{40.1^\circ}$, $r_p = 0.519$

VALUES OF Γ_L ON $G_p = 20\,dB$ CIRCLE:

$\Gamma_L = C_p + r_p e^{j\theta_1} = 0.505\angle{40.1} + 0.519 e^{j\theta_1}$

THE TABLE SHOWS TWO VALUES OF Γ_L (i.e, for $\theta_1 = \pi$ AND $\theta_1 = \frac{3\pi}{2}$), THE ASSOCIATED VALUES OF $\Gamma_s = \Gamma_{IN}^*$, Γ_{OUT}, $|\Gamma_b|$, AND $(VSWR)_{OUT}$.

Γ_L	$\Gamma_s = \Gamma_{IN}^*$	$(VSWR)_{in}$	Γ_{OUT}	$\lvert\Gamma_b\rvert$	$(VSWR)_{OUT}$
($\theta_1 = \pi$) $0.352\angle{112.2^\circ}$	$0.655\angle{163.4^\circ}$	1	$0.668\angle{-44.8^\circ}$	0.667	30.25
($\theta_1 = 3\pi/2$) $0.432\angle{-26.6^\circ}$	$0.607\angle{-179.2^\circ}$	1	$0.674\angle{-34.3^\circ}$	0.668	30.25

MAPPING OF THE $G_p = 20\,dB$ CIRCLE TO THE $\Gamma_s = \Gamma_{IN}^*$ PLANE:

$C_i = 0.823\angle{175.3^\circ}$, $r_i = 0.227$

INPUT STABILITY CIRCLE: $C_s = 1.65\angle{175.3^\circ}$, $r_s = 0.679$

THE VALUES OF $(VSWR)_{OUT} = 30.25$ SHOW THAT THE OUTPUT HAS TO BE MISMATCHED IN ORDER TO REDUCE THE GAIN TO 20 dB. THE DESIGNER CAN TRY TO REDUCE $(VSWR)_{OUT}$ BY RELAXING THE INPUT VSWR VALUE (SAY, LET $(VSWR)_{in} \le 1.5$), AS DISCUSSED IN EXAMPLE 3.8.2.

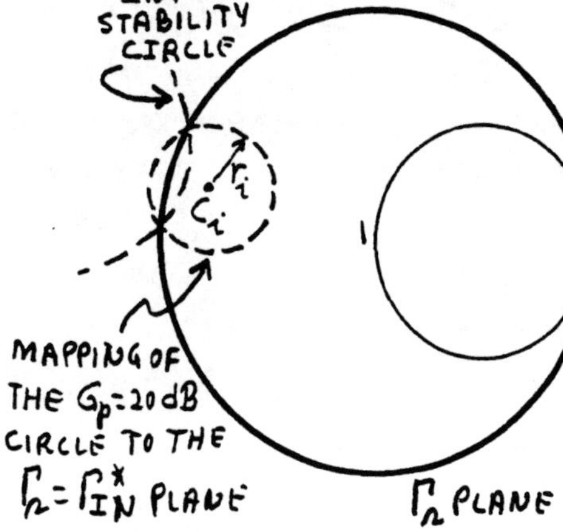

3.35) $K = 0.875$, $\Delta = 0.445\angle 160.4^\circ$ ∴ POTENTIALLY UNSTABLE

$$G_{MSG} = \frac{|S_{21}|}{|S_{12}|} = \frac{3.1}{0.125} = 24.8 \text{ OR } 13.9\,dB$$

DESIGN FOR $G_A = 10\,dB$ (i.e, 3.9 dB LESS THAN G_{MSG})

INPUT STABILITY CIRCLE:

$C_s = 3.303\angle -173.2^\circ$, $r_s = 2.392$

$G_A = 10\,dB$ CONSTANT-GAIN CIRCLE: ($g_a = 1.041$)

$C_a = 0.296\angle -173.2^\circ$, $r_a = 0.73$

VALUES OF Γ_s ON THE $G_A = 10\,dB$ CIRCLE:

$$\Gamma_s = C_a + r_a e^{j\theta_1} = 0.296\angle -173.2^\circ + 0.73 e^{j\theta_1}$$

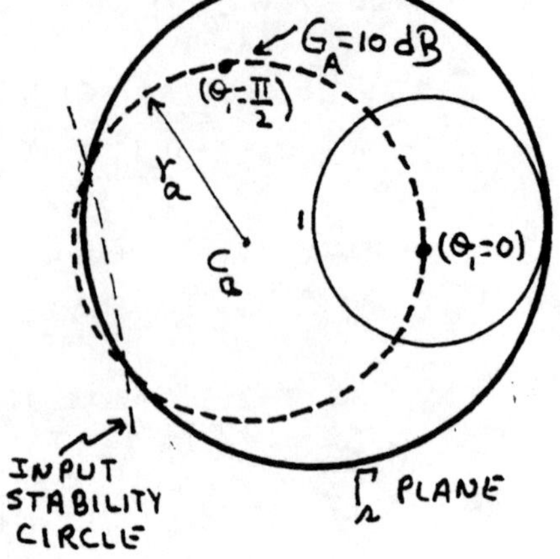

THE TABLE SHOWS TWO VALUES OF Γ_s (i.e, for $\theta_1 = 0$ AND $\theta_1 = \frac{\pi}{2}$), THE ASSOCIATED VALUES OF $\Gamma_L = \Gamma_{OUT}^*$, Γ_{IN}, $|\Gamma_a|$, AND $(VSWR)_{in}$.

Γ_s	$\Gamma_L = \Gamma_{OUT}^*$	$(VSWR)_{OUT}$	Γ_{IN}	$\lvert\Gamma_a\rvert$	$(VSWR)_{in}$
$(\theta_1 = 0)$ $0.437\angle -4.6^\circ$	$0.405\angle 70.8^\circ$	1	$0.565\angle 172^\circ$	0.802	9.1
$(\theta_1 = \pi/2)$ $0.755\angle 112.9^\circ$	$0.08\angle 133.5^\circ$	1	$0.624\angle -171.9^\circ$	0.802	9.1

MAPPING OF THE $G_A = 10\,dB$ CIRCLE TO THE $\Gamma_L = \Gamma_{OUT}^*$ PLANE:

$C_o = 0.646\angle 130.9^\circ$ AND $r_o = 0.566$

OUTPUT STABILITY CIRCLE:

$C_L = 9.33\angle -49.1^\circ$, $r_L = 10.19$

THE VALUES OF $(VSWR)_{in} = 9.1$ SHOW THAT THE INPUT HAS TO BE MISMATCHED IN ORDER TO REDUCE THE GAIN TO 10 dB. (i.e, $G_A = 10\,dB$). THE DESIGNER CAN TRY TO REDUCE $(VSWR)_{in}$ BY RELAXING THE OUTPUT VSWR.

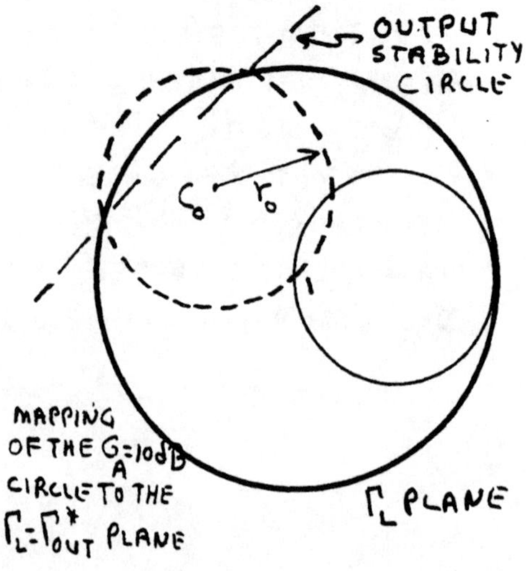

3.36) (a) dc MODEL

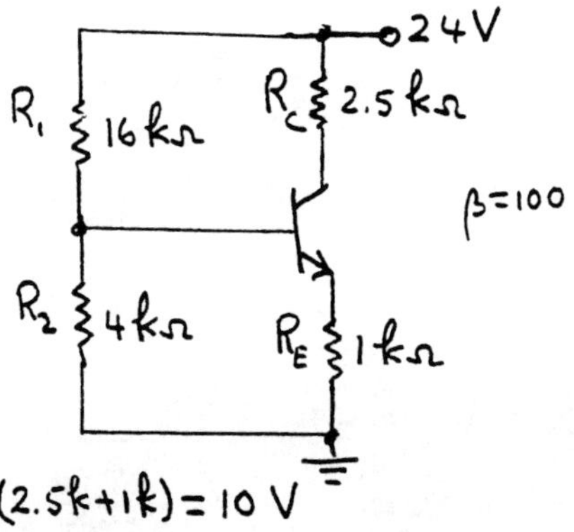

(b) $V_{TH} = \frac{24(4k)}{4k+16k} = 4.8\text{ V}$

$R_{TH} = 4k \| 16k = 3.2\text{k}\Omega$

$V_{TH} = I_B R_{TH} + 0.7 + I_C R_E \quad (I_C \approx I_E)$

$\therefore \; I_B = \frac{V_{TH} - 0.7}{R_{TH} + \beta R_E} = 40\,\mu A$

$I_C = \beta I_B = 4\text{ mA}$

$V_{CE} = V_{CC} - I_C(R_C + R_E) = 24 - 4m(2.5k + 1k) = 10\text{ V}$

(c) AT 500 MHz THE 0.1μF ($Z_c = -j0.003\,\Omega$) ACT AS SHORT CIRCUITS TO THE AC SIGNAL. THUS, RFCs ARE NOT NEEDED IN SERIES WITH THE 16kΩ AND 2.5kΩ RESISTORS.

THE 100 nH INDUCTOR ($Z_L = j314\,\Omega$) IMPEDANCE IS ABOUT 10% OF THE RESISTANCE OF $R_2 = 4k\Omega$. THUS, A RFC SHOULD BE USED IN SERIES WITH THE 4kΩ RESISTOR.

(d) ac MODEL

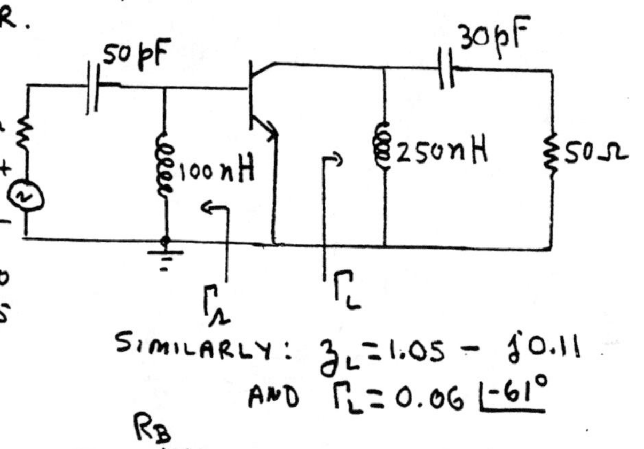

(e) $Z_{50pF} = \frac{1}{j2\pi 500\,10^6\, 50\,10^{-12}} = -j6.37\,\Omega$

$Z_{100nH} = j2\pi 500\,10^6\, 100\,10^{-9} = j314.2\,\Omega$

USING THE ZY CHART IT IS SIMPLE TO CALCULATE: $z_s = 1.015 + j0.035$

AND $\Gamma_s = 0.02\angle 65.8^\circ$

SIMILARLY: $z_L = 1.05 - j0.11$

AND $\Gamma_L = 0.06\angle{-61^\circ}$

3.37) (a) dc MODEL

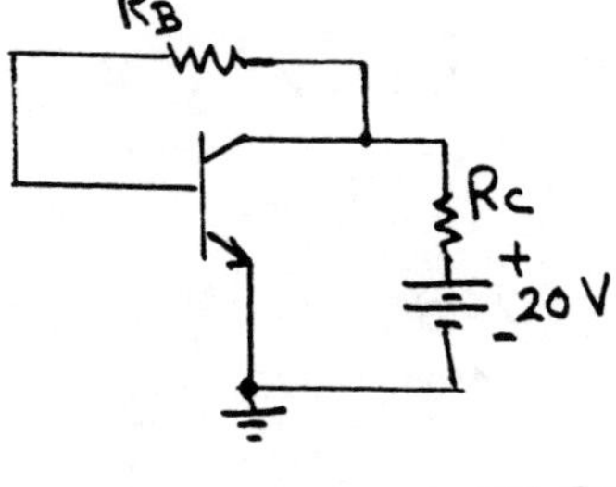

(b) $20 = I_C R_C + V_{CE}$

OR $R_C = \frac{20 - 10}{5m} = 2\text{k}\Omega$

$V_{CE} = I_B R_B + 0.7$

$R_B = \frac{10 - 0.7}{\left(\frac{5m}{100}\right)} = 186\text{ k}\Omega$

THIS TYPE OF DC BIAS RESULTS IN A LARGE VALUE FOR R_B.

(c) ac MODEL

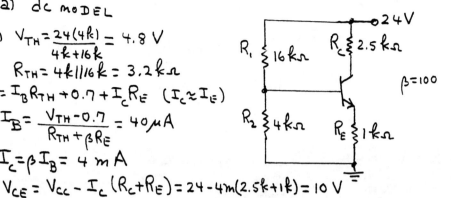

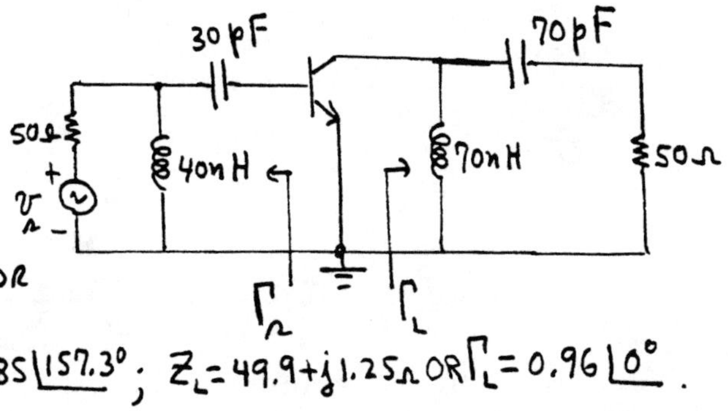

AT f=300MHz: $Z_{30pF} = -j17.68\,\Omega$,

$Z_{70pF} = -j7.58\,\Omega$, $Z_{40nH} = j75.4\,\Omega$

$Z_{70nH} = j131.9\,\Omega$

USING THE ZY SMITH CHART OR SIMPLE CALCULATIONS, WE OBTAIN:

$Z_s = 35.1 + j5.2\,\Omega$ OR $\Gamma_s = 0.185\angle 157.3^\circ$; $Z_L = 49.9 + j1.25\,\Omega$ OR $\Gamma_L = 0.96\angle 0^\circ$.

3.38) LET $V_{CC}=20V$, $R_E=\frac{10\%\,V_{CC}}{I_C}=\frac{0.1(20)}{10m}=200\,\Omega$

$$V_{CC}=V_{CE}+I_C(R_C+R_E) \Rightarrow R_C+200=\frac{20-10}{4m}=2.5k, \text{ OR } R_C=2.3\,k\Omega$$

FOR GOOD β STABILITY LET $R_{TH}=\frac{\beta R_E}{10}=\frac{100(200)}{10}=2\,k\Omega$

$$V_{TH}=I_B R_{TH}+0.75+I_C R_E=\frac{4m}{100}(2k)+0.75+4m(200)=1.63\,V$$

$$R_1=R_{TH}\frac{V_{CC}}{V_{TH}}=2.3k\,\frac{20}{1.63}=28.22\,k\Omega$$

$$R_2=\frac{R_{TH}}{1-V_{TH}/V_{CC}}=\frac{2.3k}{1-\left(\frac{1.63}{20}\right)}=2.5\,k\Omega$$

3.39) $I_{C2}=10\,mA$. LET $I_3=20\,mA$, THEN $I_{C1}=I_3-I_{C2}=20m-10m=10\,mA$

$$\therefore\ R_4=\frac{0.75}{I_{C1}}=\frac{0.75}{10m}=75\,\Omega \quad . \text{ LET } V_{CC}=20\,V$$

$$R_3=\frac{V_{CC}-V_{CE,2}}{I_3}=\frac{20-10}{20m}=500\,\Omega$$

FOR GOOD β STABILITY LET $I_{R_1}\approx I_{R_2}=20\,I_{B1}=20\,\frac{10m}{100}=2\,mA$,

$$V_{B,1}=V_{CE,2}-0.75=9.25\,V$$

$$R_1=\frac{V_{CC}-V_{B,1}}{I_{R_1}}=\frac{20-9.25}{2m}=5.37\,k\Omega$$

$$R_2=\frac{V_{B,1}}{I_{R_2}}=\frac{9.25}{2m}=4.63\,k\Omega$$

3.40) (a) IN FIG. 3.9.2a: $V_{CC}=(I_B+I_C)R_C+I_B R_B+V_{BE}$ (1)

AND $I_C=h_{FE}I_B+(h_{FE}+1)I_{CBO}$ (2)

SUBSTITUTING (2) INTO (1) GIVES:

$$V_{CC}-V_{BE}=\left(\frac{I_C}{h_{FE}}-I_{CBO}\frac{(h_{FE}+1)}{h_{FE}}\right)(R_C+R_B)+I_C R_C$$

OR

$$I_C=\frac{h_{FE}(V_{CC}-V_{BE})}{R_B+R_C(h_{FE}+1)}+\frac{I_{CBO}(h_{FE}+1)(R_C+R_B)}{R_B+R_C(h_{FE}+1)}$$

$$S_i=\frac{\partial I_C}{\partial I_{CBO}}=\frac{(h_{FE}+1)(R_C+R_B)}{R_B+R_C(h_{FE}+1)}$$

$$S_{V_{BE}}=\frac{\partial I_C}{\partial V_{BE}}=\frac{-h_{FE}}{R_B+R_C(h_{FE}+1)}$$

IN THE DERIVATION OF $S_{h_{FE}}$ WE NEGLECT THE CONTRIBUTION FROM I_{CBO}. HENCE,

$$I_{c2} = \frac{h_{FE,2}(V_{cc}-V_{BE})}{R_B + R_c(h_{FE,2}-1)} \quad \text{AND} \quad I_{c_1} = \frac{h_{FE}(V_{cc}-V_{BE})}{R_B + R_c(h_{FE}-1)}$$

$$\frac{I_{c2}}{I_{c_1}} - 1 = \frac{h_{FE,2}[R_B + R_c(h_{FE}-1)]}{h_{FE}[R_B + R_c(h_{FE,2}-1)]} - 1 = \frac{(h_{FE,2}-h_{FE})}{h_{FE}} \frac{S_{i2}}{h_{FE,2}}$$

OR $\quad S_{h_{FE}} = \dfrac{\Delta I_c}{\Delta h_{FE}} = \dfrac{I_{c2}-I_{c_1}}{h_{FE,2}-h_{FE}} = \dfrac{I_{c_1} S_{i2}}{h_{FE}\, h_{FE,2}}$

(b) $\quad I_B = \dfrac{I_c}{h_{FE}} = \dfrac{10\,10^{-3}}{50} = 0.2\,mA$, $R_B = \dfrac{V_{CE}-0.7}{I_B} = \dfrac{10-0.7}{0.2\,10^{-3}} = 46.5\,k\Omega$

$\quad R_c = \dfrac{V_{cc}-V_{CE}}{I_c + I_B} = \dfrac{20-10}{10.2\quad^3} = 980\,\Omega$

(c) IF $h_{FE} = 100$, I_c INCREASES FROM 10 mA TO:

$$I_{c2} = \frac{h_{FE,2}(V_{cc}-0.7)}{R_B + R_c(h_{FE,2}+1)} \quad \text{WHERE } h_{FE,2} = 100$$

$$= \frac{100(20-0.7)}{46.5\,10^3 + 0.98\,10^3(100+1)} = 13.3\,mA$$

AND $\quad V_{CE} = V_{cc} - I_{c2}R_c = 20 - 13.3\,10^{-3}(0.98\,10^3) = 7\,V$

3.41) LET $V_{cc} = 12\,V$, $R_E = \dfrac{10\%\,V_{cc}}{I_c} = \dfrac{0.1(12)}{1m} = 1.2\,k\Omega$
(a)

$\quad V_{cc} = V_{CE} + I_c(R_c + R_E) \Rightarrow R_c + 1.2k = \dfrac{12-6}{1m} = 6\,k\Omega$ OR $R_c = 4.8\,k\Omega$.

FROM $\quad S_i = \dfrac{(\beta+1)(R_{TH}+R_E)}{R_{TH} + (\beta+1)R_E} \Rightarrow 5 = \dfrac{101(R_{TH}+1.2k)}{R_{TH}+(101)1.2k}$ $\therefore R_{TH} = 5.05\,k\Omega$

$\quad V_{TH} = I_B R_{TH} + 0.7 + I_c R_E = \dfrac{1m}{100}(4.8k) + 0.7 + 1m(1.2k) = 1.95\,V$

$\quad R_1 = R_{TH}\dfrac{V_{cc}}{V_{TH}} = 5.05k\dfrac{12}{1.95} = 31.1\,k\Omega$

$\quad R_2 = \dfrac{R_{TH}}{1 - V_{TH}/V_{cc}} = \dfrac{5.05k}{1 - 1.95/12} = 6.03\,k$

(b) $I_{CBO,2} = 1\,10^{-6}\,2^{\frac{75-25}{10}} = 32\,\mu A$, $h_{FE,2} = 100 + \dfrac{0.5}{100}(75-25)100 = 125$

$\quad S_i = 5$, $S_{V_{BE}} = \dfrac{-100}{5.05\,10^3 + (101)1.2\,10^3} = 0.8\,10^{-3}$, $S_{h_{FE}} = \dfrac{10^{-3}(5.4)}{100(125)} = 4.3\,10^{-7}$

AT 75°C : $\quad I_c = \dfrac{125(1.95-0.7)}{5.05\,10^3 + (126)1.2\,10^3} + \dfrac{126(32\,10^{-6})(5.05\,10^3 + 1.2\,10^3)}{5.05\,10^3 + (126)1.2\,10^3}$

$\quad\quad\quad = 1.16\,mA$

3.42) $I_D = I_{DSS}\left[1 - \frac{V_{GS}}{V_P}\right]^2 \Rightarrow 10m = 30m\left[1 + \frac{V_{GS}}{3}\right]^2 \quad \therefore V_{GS} = -1.268V$

$R_s = \frac{-V_{GS}}{I_D} = \frac{1.268}{10m} = 126.8\,\Omega$

LET $V_{CC} = V_{EE} = 10V$ AND $I_{R_1} = 20mA$, WHERE $I_{R_1} = I_C + I_D$

$\therefore I_C = 20m - 10m = 10\,mA$

THE VOLTAGE AT THE GATE IS ZERO (i.e, ACROSS R_G). HENCE,

$$R_S = \frac{0 - (-V_{EE})}{I_C} = \frac{10}{10m} = 1\,k\Omega$$

$$V_E = V_{DS} + V_{R_S} = 3 + 1.268 = 4.27\,V$$

THEN, $R_1 = \frac{V_{CC} - V_E}{I_{R_1}} = \frac{10 - 4.27}{20m} = 287\,\Omega$

LET $I_{R_2} \approx I_{R_3} = 20\,I_B = 20\,\frac{10m}{100} = 2\,mA$

$$V_B = V_E - 0.7 = 4.27 - 0.7 = 3.57\,V$$

$$R_2 = \frac{V_{CC} - V_B}{I_{R_2}} = \frac{10 - 3.57}{2m} = 3.21\,k\Omega$$

$$R_3 = \frac{V_B}{I_{R_3}} = \frac{3.57}{2m} = 1.78\,k\Omega$$

4.1) (a)

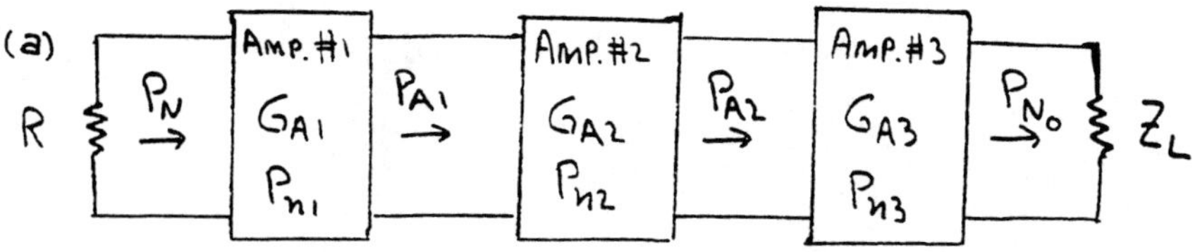

$P_N = kTB \qquad P_{A1} = G_{A1}P_N + P_{n1} \qquad P_{A2} = G_{A2}P_{A1} + P_{n2} \qquad P_{N_o} = G_{A3}P_{A2} + P_{n3}$

$$F = \frac{P_{N_o}}{P_N G_{A1} G_{A2} G_{A3}} = \frac{G_{A3}G_{A2}G_{A1}P_N + G_{A3}G_{A2}P_{n1} + G_{A3}P_{n2} + P_{n3}}{P_N G_{A1} G_{A2} G_{A3}}$$

$$= 1 + \frac{P_{n1}}{P_N G_{A1}} + \frac{P_{n2}}{P_N G_{A1} G_{A2}} + \frac{P_{n3}}{P_N G_{A1} G_{A2} G_{A3}} \qquad (1)$$

LET: $F_1 = 1 + \frac{P_{n1}}{P_N G_{A1}}$, $F_2 = 1 + \frac{P_{n2}}{P_N G_{A2}}$, $F_3 = 1 + \frac{P_{n3}}{P_N G_{A3}}$

THEREFORE, (1) CAN BE EXPRESSED IN THE FORM:

$$F = F_1 + \frac{F_2 - 1}{G_{A1}} + \frac{F_3 - 1}{G_{A1}G_{A2}}$$

(b) $F = 10^{0.1} + \frac{10^{0.3} - 1}{10^1} = 1.358$ OR 1.33 dB

4.2) (a) THE GAIN-CIRCLES ARE THE CONSTANT-GAIN G_A CIRCLES.

FROM FIG. P.2(a): $F_{min} = 3.3$ dB, $\Gamma_{opt} \approx 0.56\angle{-155^\circ}$.

r_n IS EVALUATED BY READING A VALUE OF F AND ITS ASSOCIATED Γ_s, AND USING (4.3.4). FROM FIG. P.2(a) WE READ:

$F = 4.5$ dB AT $\Gamma_s \approx 0.56\angle{-94^\circ}$. THEN

$$10^{0.45} = 10^{0.33} + \frac{4r_n \left|0.56\angle{-94^\circ} - 0.56\angle{-155^\circ}\right|^2}{(1-(0.56)^2)\left|1 + 0.56\angle{-155^\circ}\right|^2} \Rightarrow r_n = 0.108$$

FROM FIG. P.2(b): $F_{min} = 1.7$ dB, $\Gamma_{opt} \approx 0.215\angle{149^\circ}$.

USING: $F = 3$ dB AT $\Gamma_s = 0.445\angle{26^\circ}$ WE OBTAIN $r_n = 0.202$

(b) FOR THE $G_A = 10.7$ dB CIRCLE IN FIG. 4.3.4:

$g_a = 4.158$, $C_a = 0.574\angle{-154^\circ}$, $r_a = 0.423$

4.3) (a) THE $G_A = 14$ dB CIRCLE AND THE $F = 2$ dB CIRCLE INTERSECT AT TWO POINTS. THE VALUE OF Γ_s AT THESE THE POINTS ARE:

$\Gamma_s \approx 0.5\angle{160^\circ}$ AND $\Gamma_s \approx 0.25\angle{-150^\circ}$

(b) LET $\Gamma_s = 0.5\angle{160^\circ}$, THEN $\Gamma_{OUT} = 0.657\angle{-73.3^\circ}$

FOR $(VSWR)_{out} = 1$: $\Gamma_L = \Gamma^*_{OUT} = 0.657\angle{73.3^\circ}$

THEN, $\Gamma_{IN} = 0.8\angle{165.6^\circ}$, $|\Gamma_a| = 0.678$, $(VSWR)_{in} = 5.2$

4.4) (a) $K = 2.25$ AND $\Delta = 0.246\angle 112.8°$ $\therefore$ UNCONDITIONALLY STABLE

(b) $G_{A,max} = \frac{|S_{21}|}{|S_{12}|}\left(K - \sqrt{K^2-1}\right) = 9.36$ OR 9.71 dB

(c) $G_A = 9.71 - 3 = 6.71$ dB

FOR THE $G_A = 6.71$ dB CIRCLE: $g_a = 1.173$,

$C_a = 0.42\angle 174.5°$, $r_a = 0.515$

(d) FOR THE 3 dB NOISE CIRCLE:

$C_{F_i} = 0.405\angle 145°$, $r_{F_i} = 0.388$

FOR THE 4 dB NOISE CIRCLES:

$C_{F_i} = 0.279\angle 145°$, $r_{F_i} = 0.616$

THE $F_i = 4$ dB CIRCLE IS DRAWN ON THE SMITH CHART.

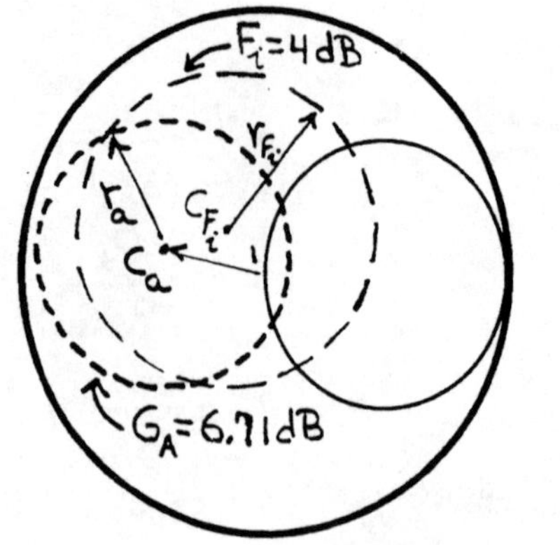

Γ_s PLANE

(e) FOR $G_{A,max}$: $\Gamma_s = \Gamma_{Ms} = 0.667\angle 174.5°$, $\Gamma_L = \Gamma_{ML} = 0.587\angle 102.2°$.

$$\therefore F = 10^{0.25} + \frac{4\left(\frac{5}{50}\right)\left|0.667\angle 174.5° - 0.5\angle 145°\right|^2}{\left(1-(0.667)^2\right)\left|1+0.5\angle 145°\right|^2} = 1.97 \text{ OR } 2.95 \text{ dB}$$

4.5) $F_{min} = 3.3$ dB, $\Gamma_{opt} = 0.56\angle -155°$, $r_n = 0.108$

$K = 1.1$, $\Delta = 0.273\angle 126.8°$ $\therefore$ UNCONDITIONALLY STABLE

DESIGN WITH $\Gamma_s = \Gamma_{opt} = 0.56\angle -155°$ AND $\Gamma_L = \Gamma_{OUT}^* = 0.686\angle 59.4°$

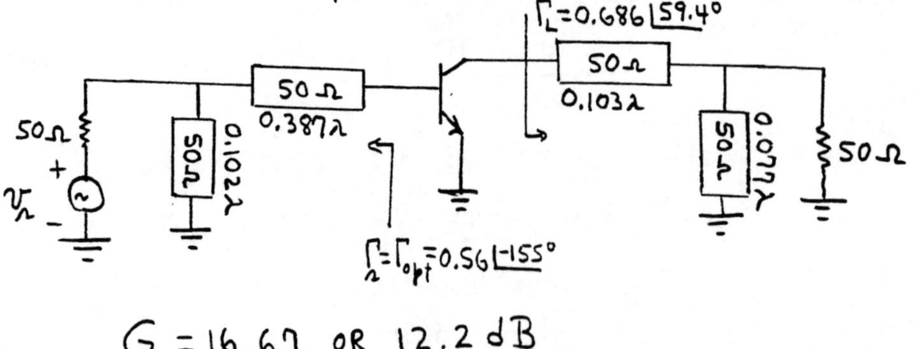

$G_A = 16.67$ OR 12.2 dB

$G_T = G_A = 16.67$ OR 12.2 dB

$G_p = 20.62$ OR 13.1 dB

ALSO: $\Gamma_{IN} = 0.801\angle 156.2°$, $|\Gamma_a| = 0.438$, $(VSWR)_{in} = 2.6$, $(VSWR)_{OUT} = 1$.

4.6) $K = 2.18$, $\Delta = 0.555\angle{-178.3^\circ}$ $\therefore$ UNCONDITIONALLY STABLE

DESIGN WITH $\Gamma_s = \Gamma_{opt} = 0.65\angle{120^\circ}$ AND $\Gamma_L = \Gamma_{OUT}^* = 0.804\angle{67^\circ}$

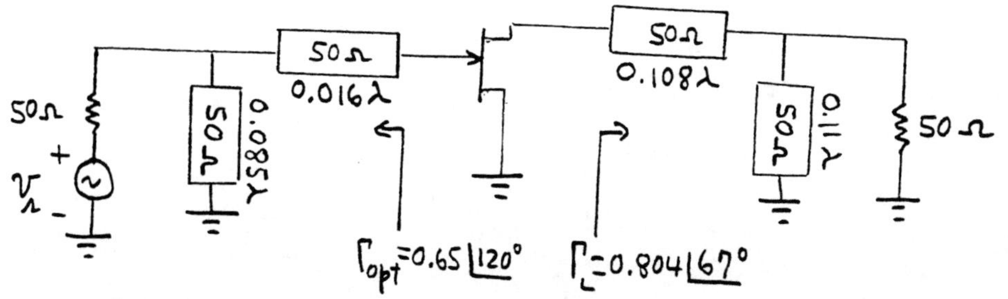

$G_A = G_T = 75.6$ OR 18.8 dB

ALSO: $\Gamma_{IN} = 0.806\angle{-119.6^\circ}$, $|\Gamma_a| = 0.327$, $(VSWR)_{in} = 1.97$, $(VSWR)_{OUT} = 1$

4.7) $K = 1.235$, $\Delta = 0.175\angle{166.2^\circ}$ $\therefore$ UNCONDITIONALLY STABLE

$G_{A,max} = 43.07$ OR 16.3 dB, $\Gamma_{Ms} = 0.787\angle{-170.1^\circ}$, $\Gamma_{ML} = 0.749\angle{44.8^\circ}$

Γ_s	$\Gamma_L = \Gamma_{OUT}^*$	Γ_{IN}	$\lvert\Gamma_a\rvert$	$(VSWR)_{in}$	G_A(dB)	F(dB)
$0.2\angle{155^\circ}$	$0.528\angle{41.7^\circ}$	$0.704\angle{170.8^\circ}$	0.62	4.27	13.75	1.6
$0.32\angle{172^\circ}$	$0.559\angle{42.9^\circ}$	$0.715\angle{170.8^\circ}$	0.535	3.30	14.5	1.66
$0.43\angle{180^\circ}$	$0.592\angle{43.7^\circ}$	$0.726\angle{170.7^\circ}$	0.446	2.61	15.1	1.82
$0.59\angle{-174^\circ}$	$0.650\angle{44.5^\circ}$	$0.747\angle{170.6^\circ}$	0.290	1.81	15.8	2.30
$0.787\angle{-170.1^\circ}$	$0.749\angle{44.8^\circ}$	$0.787\angle{170.1^\circ}$	0	1	16.3	3.76

THE DESIGN CAN BE PERFORMED WITH $\Gamma_s = 0.59\angle{-174^\circ}$ AND $\Gamma_L = 0.65\angle{44.5^\circ}$.

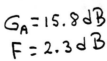

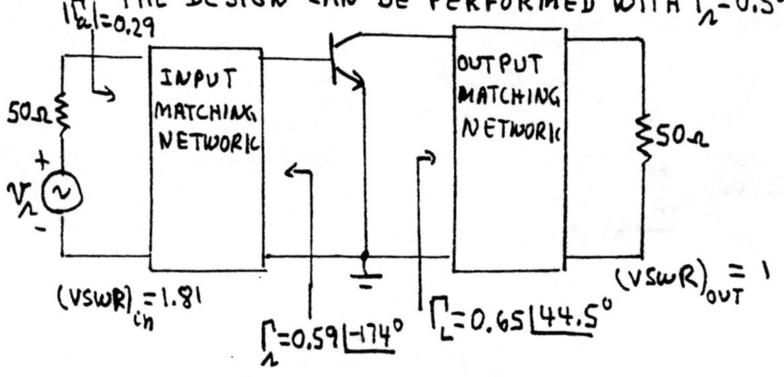

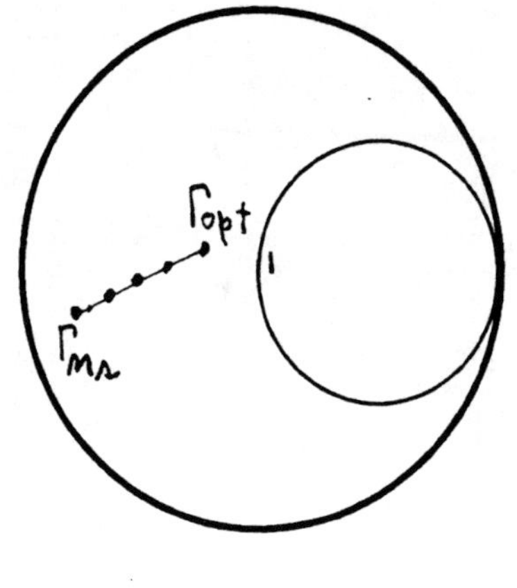

$G_A = 15.8$ dB
$F = 2.3$ dB

4.8) $K=0.96$, $\Delta = 0.6\angle -73.1°$ ∴ POTENTIALLY UNSTABLE

INPUT STABILITY CIRCLE: $C_s = 1.34\angle 62.7°$, $r_s = 0.345$

OUTPUT STABILITY CIRCLE: $C_L = 1.55\angle 47.2°$, $r_L = 0.56$

DESIGN WITH $\Gamma_s = \Gamma_{opt} = 0.73\angle 60°$ AND $\Gamma_L = \Gamma_{OUT}^* = 0.787\angle 42.2°$

BOTH Γ_{opt} AND Γ_L ARE IN THE STABLE REGION (AS EXPECTED).

A DESIGN FOR Γ_{opt} AND Γ_L IS SHOWN BELOW:

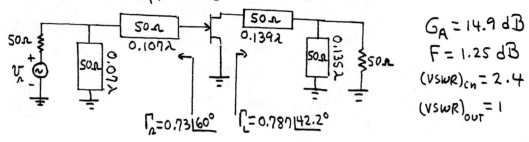

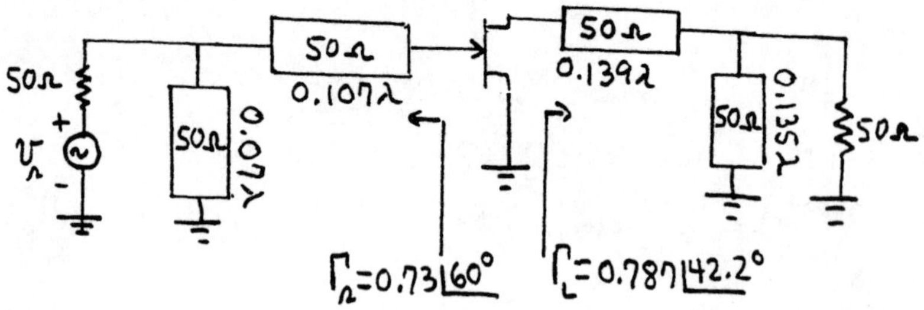

$G_A = 14.9$ dB

$F = 1.25$ dB

$(VSWR)_{in} = 2.4$

$(VSWR)_{OUT} = 1$

4.9) THIS TRANSISTOR IS POTENTIALLY UNSTABLE.

$G_{MSG} = 59.6$ OR 17.8 dB. LET US DESIGN FOR $G_A = 14$ dB.

FIG. P4.2(b) SHOWS THAT THE $G_A = 14$ dB CIRCLE AND THE $F = 2$ dB CIRCLE INTERSECT AT TWO POINTS, NAMELY:

$\Gamma_s \approx 0.5\angle 160°$ AND $\Gamma_s \approx 0.25\angle -150°$

Γ_s	G_A(dB)	F(dB)	$\Gamma_L = \Gamma_{OUT}^*$	$(VSWR)_{OUT}$	Γ_{IN}	$\lvert\Gamma_a\rvert$	$(VSWR)_{in}$
$0.5\angle 160°$	13.6	2	$0.657\angle 73.3°$	1	$0.8\angle 165.6°$	0.678	5.2
$0.25\angle -150°$	13.5	1.9	$0.685\angle 65.3°$	1	$0.793\angle 162.4°$	0.683	5.3

SELECT $\Gamma_s = 0.5\angle 160°$ AND $\Gamma_L = 0.657\angle 73.3°$.

$G_T = G_A = 13.6$ dB AND $G_P = 17.2$ dB

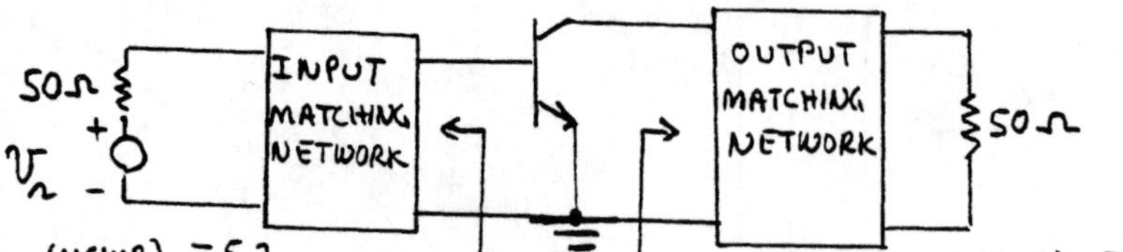

4.10) TRANSISTOR #1: $K=0.869$, $\Delta=0.348\angle{-39.2°}$ ∴ POTENTIALLY UNSTABLE

$G_{MSG}=44$ OR $16.4\,dB$.

INPUT STABILITY CIRCLE: $C_s=16.75\angle{-37.3°}$, $r_s=17.61$

OUTPUT STABILITY CIRCLE: $C_L=4.06\angle{44.5°}$, $r_L=3.16$

THE $G_A=14\,dB$, $12\,dB$, AND $10\,dB$ CIRCLES, THE $F_i=2\,dB$, AND $2.5\,dB$ CIRCLES, AND THE INPUT STABILITY CIRCLE ARE DRAWN IN THE SMITH CHART.

G_A(dB)	C_a	r_a
14	$0.561\angle{142.7°}$	0.597
12	$0.35\angle{142.7°}$	0.724
10	$0.219\angle{142.7°}$	0.821

F_i(dB)	C_{F_i}	r_{F_i}
2	$0.306\angle{77°}$	0.258
2.5	$0.254\angle{77°}$	0.458

LET US TRY A DESIGN WITH Γ_s AT POINT "a": $\Gamma_s=0.08\angle{31°}$. WITH THIS VALUE OF Γ_s WE FIND: $\Gamma_L=\Gamma_{OUT}^*=0.485\angle{37.5°}$, $\Gamma_{IN}=0.555\angle{-140°}$, $|\Gamma_a|=0.577$, AND $(VSWR)_{in}=3.7$. THE VALUE OF $(VSWR)_{in}$ CAN BE IMPROVED IF WE DESIGN FOR A HIGHER $(VSWR)_{out}$.

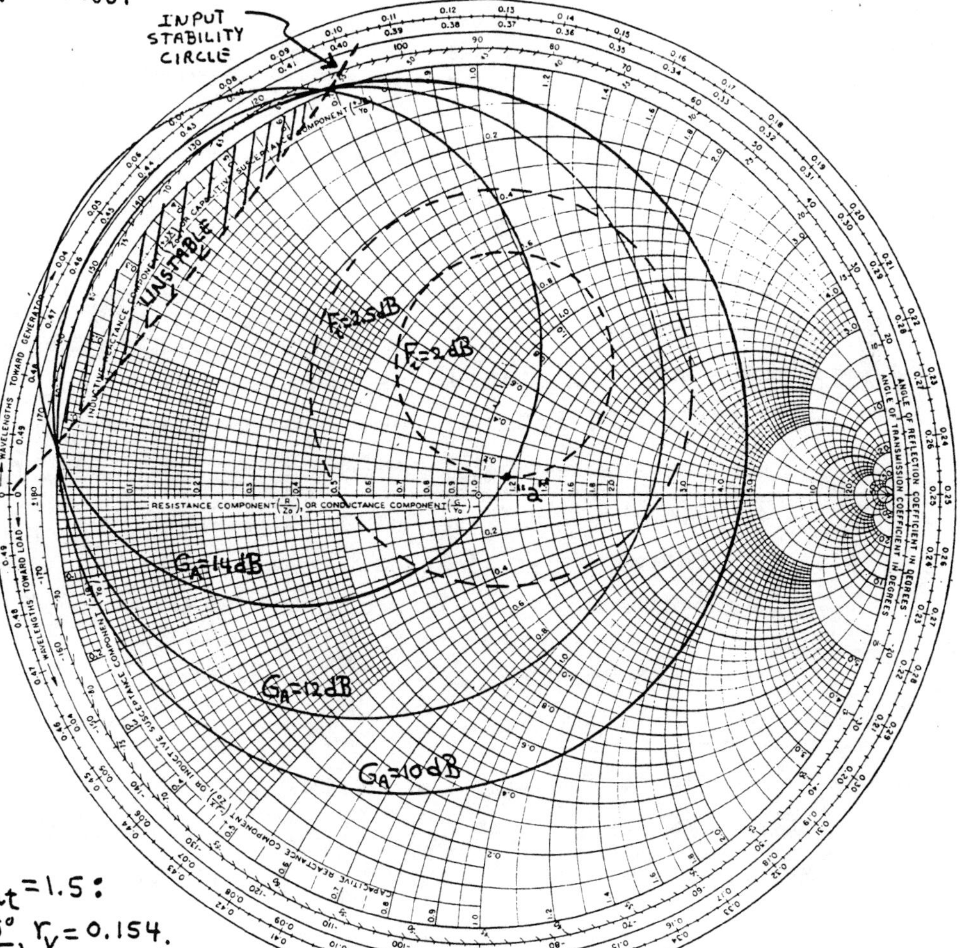

FOR $(VSWR)_{out}=1.5$:

$C_{V_o}=0.47\angle{37.5°}$, $r_{V_o}=0.154$.

SELECTING $\Gamma_L=0.32\angle{37.5°}$ IT FOLLOWS THAT

$\Gamma_{IN}=0.449\angle{-134.7°}$, $|\Gamma_a|=0.47$, AND $(VSWR)_{in}=2.78$

TRANSISTOR #2 : THE DESIGN PROCEDURE IS SIMILAR TO THE ONE USED WITH TRANSISTOR #1. ONLY THE CALCULATIONS FOR THE G_A CIRCLES AND FOR THE NOISE CIRCLES ARE GIVEN.

$K = 0.244$, $\Delta = 0.263\angle 56.2^\circ$ ∴ POTENTIALLY UNSTABLE

$G_{MSG} = 22.5$ dB

INPUT STABILITY CIRCLE: $C_s = 2.04\angle 172.2^\circ$, $r_s = 1.55$

OUTPUT STABILITY CIRCLE: $C_L = 1.63\angle 21.9^\circ$, $r_L = 1.07$

THE CALCULATIONS FOR THE $G_A = 18$ dB AND 16 dB CIRCLES, AND FOR THE $F_i = 1.2$ dB AND 1.5 dB CIRCLES ARE:

G_A(dB)	C_a	r_a
18	$0.376\angle 172.2^\circ$	0.796
16	$0.255\angle 172.2^\circ$	0.849

F_i(dB)	C_{F_i}	r_{F_i}
1.2	$0.39\angle 32^\circ$	0.246
1.5	$0.35\angle 32^\circ$	0.378

4.11) $K = 0.774$, $\Delta = 0.46\angle -129.3^\circ$ ∴ POTENTIALLY UNSTABLE.

$G_{MSG} = 14.36$ dB

INPUT STABILITY CIRCLE $\begin{cases} C_s = 2.06\angle 120.3^\circ \\ r_s = 1.19 \end{cases}$ OUTPUT STABILITY CIRCLE $\begin{cases} C_L = 334.6\angle -65.4^\circ \\ r_L = 335.4 \end{cases}$

CONSIDER THE $G_p = 10$ dB AND 12 dB CIRCLES:

G_p(dB)	C_p	r_p
12	$0.58\angle 114.6^\circ$	0.663
10	$0.366\angle 114.6^\circ$	0.754

THE TRANSFORMATIONS OF THE $G_p = 12$ dB AND 10 dB CIRCLES TO THE $\Gamma_s = \Gamma_{IN}^*$ PLANE ARE SHOWN IN THE SMITH CHART. ALSO, THE $F_i = 1.2$ dB AND 1.5 dB CIRCLES ARE SHOWN, AS WELL AS ONE $(VSWR)_{in} = 2$ CIRCLE.

$G_p = 12$ dB CIRCLE IN THE $\Gamma_s = \Gamma_{IN}^*$ PLANE: $C_i = 0.959\angle 120.3^\circ$, $r_i = 0.366$

$G_p = 10$ dB CIRCLE IN THE $\Gamma_s = \Gamma_{IN}^*$ PLANE: $C_i = 0.858\angle 120.3^\circ$, $r_i = 0.372$

$F_i = 1.2$ dB CIRCLE IN THE Γ_s PLANE: $C_{F_i} = 0.654\angle 55^\circ$, $r_{F_i} = 0.188$

$F_i = 1.5$ dB CIRCLE IN THE Γ_s PLANE: $C_{F_i} = 0.621\angle 55^\circ$, $r_{F_i} = 0.252$

A POINT ON THE $G_p = 10$ dB CIRCLE IS $\Gamma_L = 0.416\angle -57.3^\circ$. THEN, $\Gamma_{IN} = 0.62\angle -97^\circ$. THE VALUE $\Gamma_s = \Gamma_{IN}^*$ RESULTS IN $(VSWR)_{in} = 1$, AND A VERY-HIGH NOISE FIGURE. THE CIRCLE $(VSWR)_{in} = 2$ IS DRAWN (i.e, $C_{V_i} = 0.576\angle 97^\circ$ AND $r_{V_i} = 0.214$). A CONVENIENT VALUE OF Γ_s IS AT POINT "a": $\Gamma_s = 0.56\angle 79^\circ$, AND IT FOLLOWS THAT

Γ_s	Γ_L	G_p(dB)	F_i(dB)	Γ_{OUT}	$\lvert\Gamma_a\rvert$	$(VSWR)_{in}$	$\lvert\Gamma_b\rvert$	$(VSWR)_{out}$
$0.56\angle 79^\circ$	$0.416\angle -57.3^\circ$	10	1.5	$0.35\angle -107^\circ$	0.286	1.8	0.664	4.9

IMPEDANCE OR ADMITTANCE COORDINATES

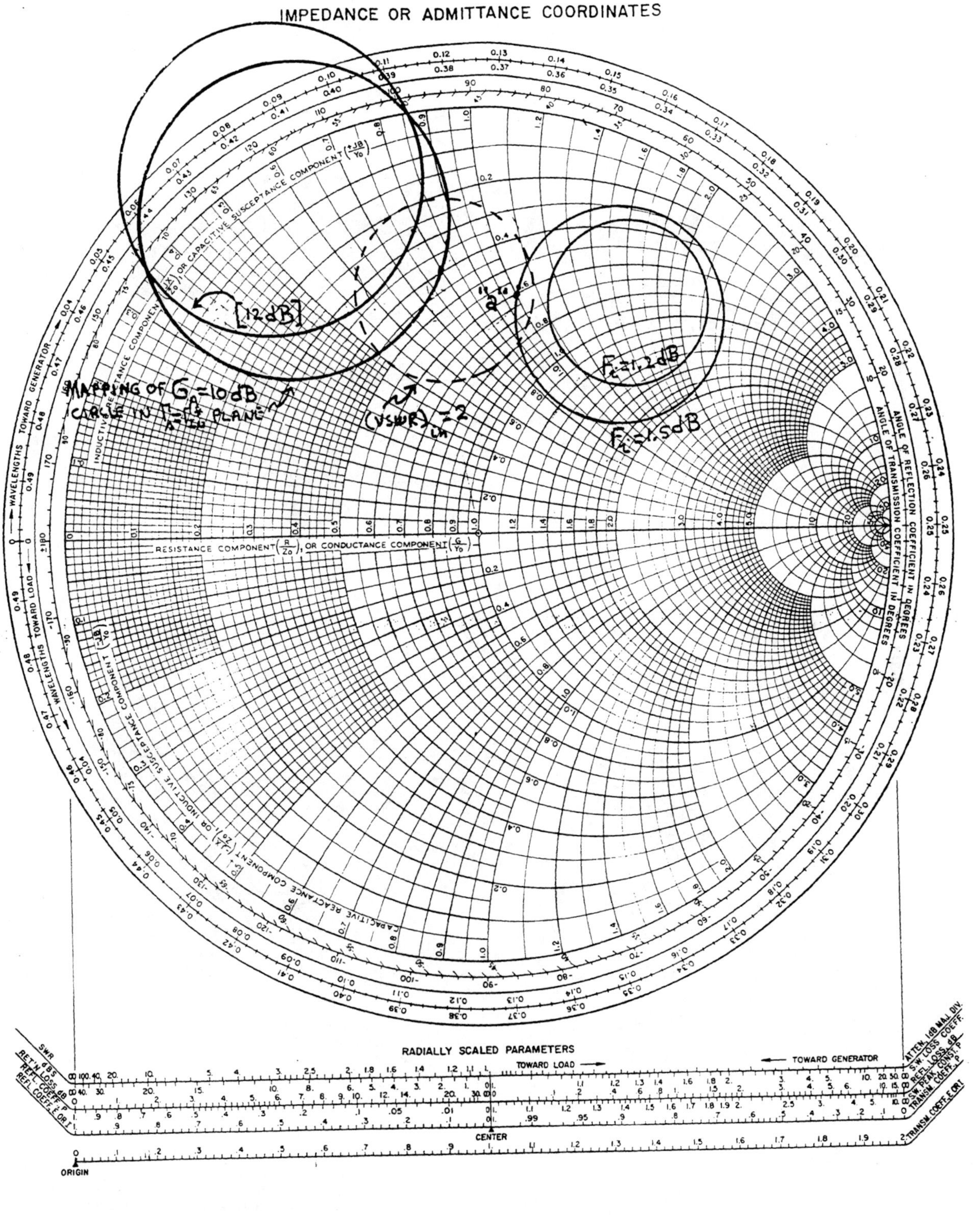

4.12) $K = 1.04$, $\Delta = 0.83\angle{-47.5^\circ}$ $\therefore$ UNCONDITIONALLY STABLE

$G_{p,max} = 5.166$ OR $7.13\,dB$. CONSIDER THE $G_p = 6\,dB$ AND $5\,dB$ CIRCLES.

$G_p = 6\,dB$ CIRCLE IN THE $\Gamma_s = \Gamma_{IN}^*$ PLANE : $C_i = 0.257\angle{-10^\circ}$, $r_i = 0.443$

$G_p = 5\,dB$ CIRCLE IN THE $\Gamma_s = \Gamma_{IN}^*$ PLANE : $C_i = 0.223\angle{-10^\circ}$, $r_i = 0.541$

$F_i = 2.1\,dB$ CIRCLE IN THE Γ_s PLANE : $C_{F_i} = 0.569\angle{71^\circ}$, $r_{F_i} = 0.185$

$F_i = 2.3\,dB$ CIRCLE IN THE Γ_s PLANE : $C_{F_i} = 0.539\angle{71^\circ}$, $r_{F_i} = 0.262$

THESE CIRCLES ARE DRAWN IN THE SMITH CHART. SELECTING Γ_s AT POINT "a" RESULTS IN $F_i < 2.1\,dB$, $G_p = 6\,dB$, AND $(VSWR)_{in} = 1$.

THAT IS:

Γ_s	Γ_L	Γ_{IN}	Γ_{OUT}	G_p(dB)	F_i(dB)	$(VSWR)_{in}$	$\lvert\Gamma_b\rvert$	$(VSWR)_{OUT}$
$0.44\angle{63^\circ}$	$0.86\angle{50.5^\circ}$	$0.44\angle{-63^\circ}$	$0.765\angle{-44.7^\circ}$	6	2	1	0.355	2.1

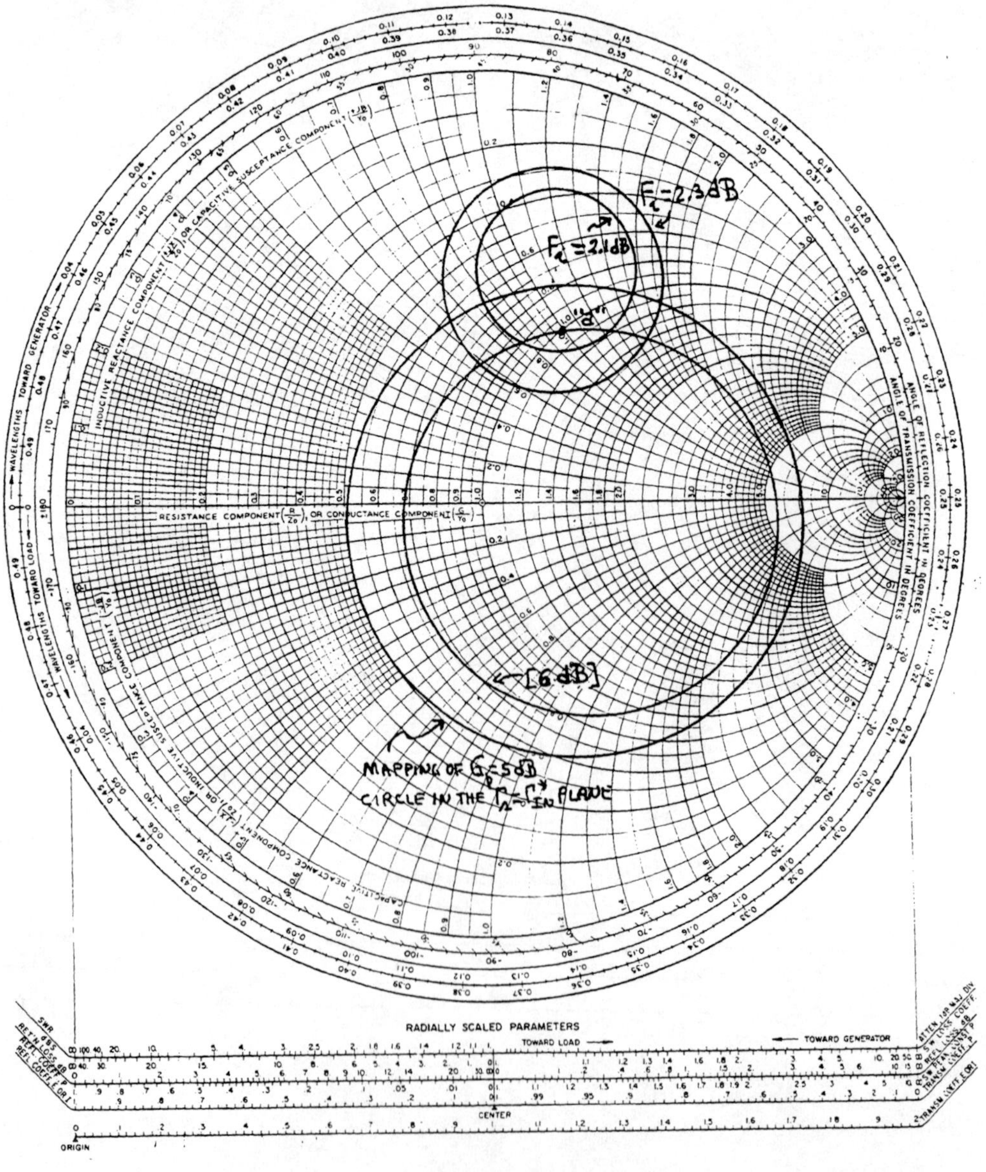

4.13) $G_A(dB) = 14 + 14 + 14 - 7.5 + 16 + 16 = 66.5$ dB

THE NOISE FIGURE OF THE LNB IS DETERMINED BY THE NOISE FIGURE OF THE LNA (i.e., THE FIRST 3 AMPLIFIERS).

$$F = F_1 + \frac{F_2 - 1}{G_{A1}} + \frac{F_3 - 1}{G_{A1}G_{A2}} = 10^{0.05} + \frac{10^{0.09} - 1}{10^{1.4}} + \frac{10^{0.11} - 1}{10^{1.4}10^{1.4}} = 1.131 \text{ OR } 0.54\,dB$$

THE CONTRIBUTION TO F FROM THE MIXER, AMP. 5, AND AMP. 6 IS NEGLIGIBLE.

4.14) THE FOLLOWING CALCULATIONS ARE MADE:

f(MHz)	$\lvert S_{21}\rvert^2$	$G_{s,max}$	$G_{L,max}$	$G_{TU,max}$
150	25 (14dB)	0.45 dB	7.6 dB	22 dB
250	16 (12dB)	0.38 dB	5.8 dB	18.2 dB
400	7.94 (9dB)	0.28 dB	4.6 dB	13.9 dB

OBSERVE THAT $G_{s,max}$ IS VERY SMALL AT THE THREE FREQUENCIES OF INTEREST. HENCE, ONLY THE LOAD MATCHING NETWORK WILL BE USED TO COMPENSATE FOR THE VARIATIONS IN $\lvert S_{21}\rvert^2$. THE INPUT MATCHING NETWORK CAN BE DESIGNED TO PROVIDE A GOOD $(VSWR)_{in}$, OR FOR $G_s = 0$ dB.

FOR $G_{TU} = 12$ dB, DESIGN G_L TO HAVE -2 dB AT 150 MHz, 0 dB AT 250 MHz, AND 3 dB AT 400 MHz.

f(MHz)	G_L(dB)	g_L	C_{g_L}	r_{g_L}
150	−2	0.109	$0.378\angle 16^\circ$	0.62
250	0	0.260	$0.494\angle 15^\circ$	0.494
400	3	0.687	$0.7\angle 26^\circ$	0.242

THE GAIN CIRCLES ARE DRAWN IN THE SMITH CHART. AFTER SOME TRIALS AND ERRORS, THE FOLLOWING MATCHING CIRCUIT WAS OBTAINED:

$Z_{L_1} = j\omega L_1$

$Y_{L_2} = -j/\omega L_2$

$Z_{L_3} = j\omega L_3$

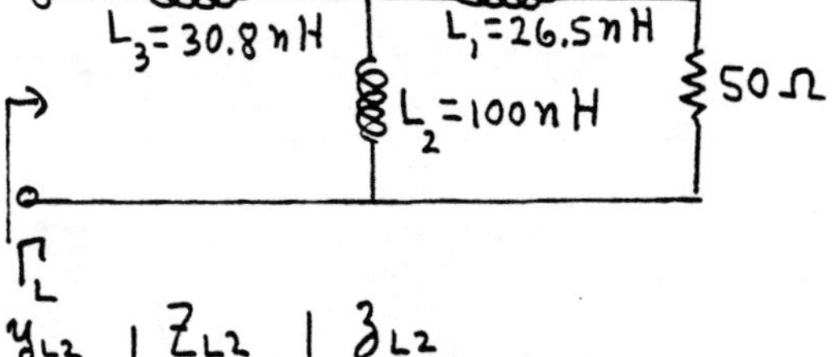

f(MHz)	Z_{L1}	$z_{L1} = \frac{Z_{L1}}{50}$	Y_{L2}	y_{L2}	Z_{L2}	z_{L2}
150	$j24.97$	$j0.5$	$-j10.61$mS	$-j0.53$	$j29.03$	$j0.58$
250	$j41.62$	$j0.83$	$-j6.36$mS	$-j0.32$	$j48.4$	$j0.97$
400	$j66.6$	$j1.33$	$-j3.98$mS	$-j2$	$j77.4$	$j1.55$

NORMALIZED IMPEDANCE AND ADMITTANCE COORDINATES

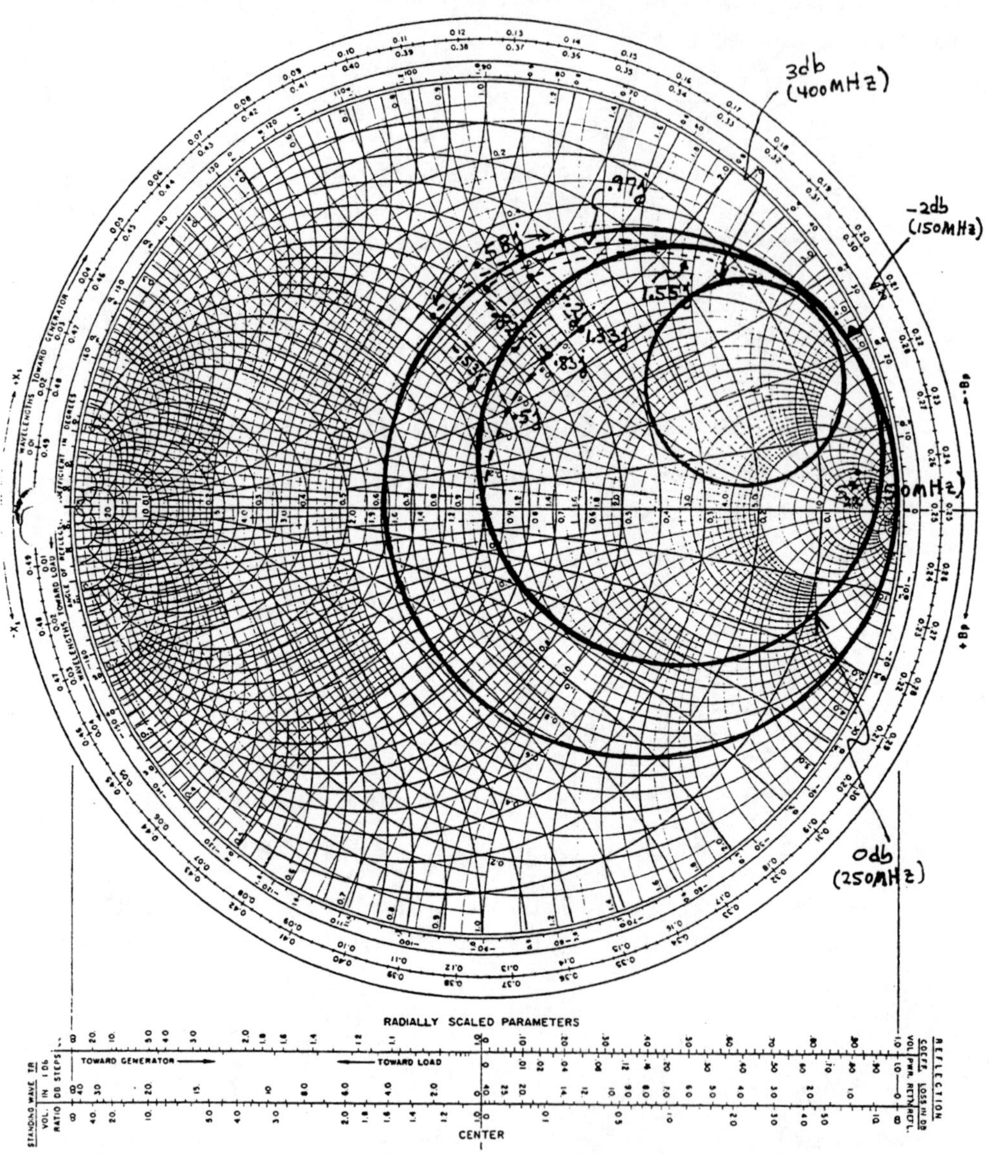

4.15) $G_{TU} = G_s G_o G_L$, $G_{TU}(dB) = G_s(dB) + G_o(dB) + G_L(dB)$

f(GHz)	$G_{s,max}$(dB)	$G_o = \lvert S_{21}\rvert^2$(dB)	$G_{L,max}$(dB)	$G_{TU,max}$(dB)
1	2.3	14.05	4.25	20.6
2	1.86	9.97	3.6	15.43

NOISE CIRCLES CALCULATIONS:

f(GHz)	F_i(dB)	C_{F_i}	r_{F_i}
1	2	$0.395\angle 23°$	0.378
	3	$0.286\angle 23°$	0.591
	4	$0.212\angle 23°$	0.708
2	2	$0.377\angle 88°$	0.261
	3	$0.304\angle 88°$	0.477
	4	$0.244\angle 88°$	0.604

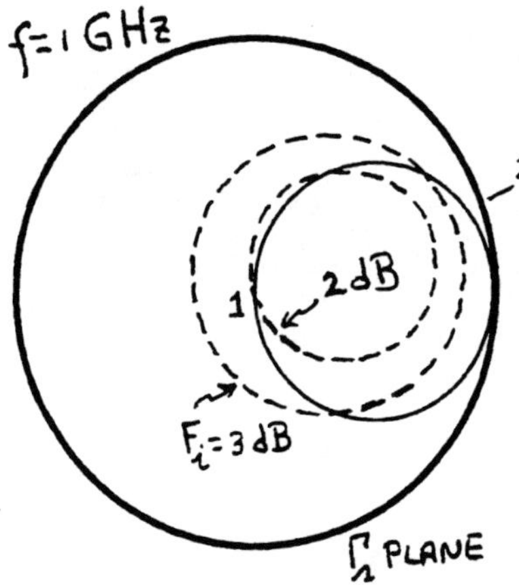

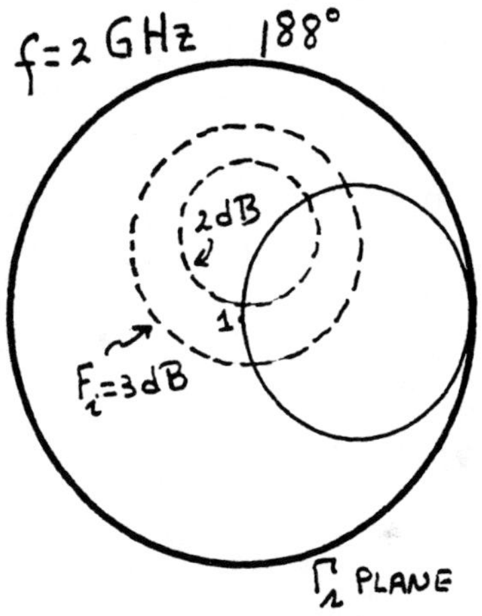

IT IS SEEN THAT THE NOISE FIGURE IS LESS THAN 3 dB AT 1 GHz AND 2 GHz WITH $\Gamma_s = 0$. THE INPUT MATCHING NETWORK IS DESIGNED WITH $\Gamma_s = 0$, AND IT FOLLOWS THAT $G_s = 0$ dB.

A DESIGN FOR $G_{TU} = 10$ dB REQUIRES THAT $G_L = -4$ dB AT 1 GHz, AND $G_L = 0$ dB AT 2GHz. THAT IS,

AT 1GHz: $G_{TU} = 0 + 14.05 - 4 \approx 10$dB ; AT 2GHz: $G_{TU} = 0 + 9.97 + 0 \approx 10$ dB

A MATCHING CIRCUIT FOR THIS DESIGN IS:

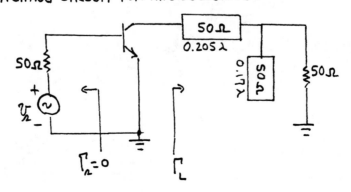

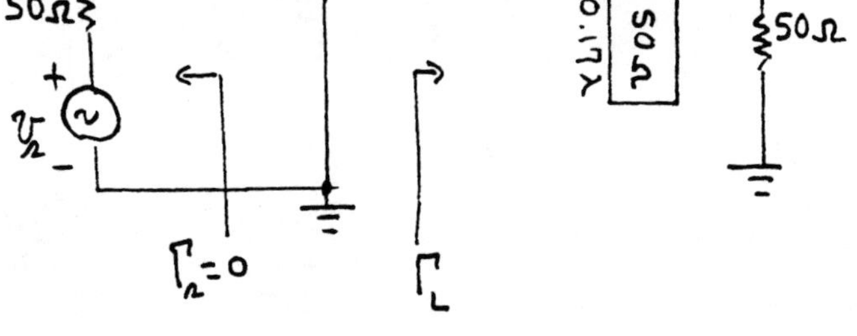

4.16) (a) $S_{11a}=S_{11b}=S_{11}$ $\qquad$ $S_{21a}=S_{21b}=S_{21}$

$S_{12a}=S_{12b}=S_{12}$ $\qquad$ $S_{22a}=S_{22b}=S_{22}$

$$S_{11}=\frac{e^{-j\pi}}{2}(S_{11a}-S_{11b})=0 \quad ,(VSWR)_{in}=1$$

$$S_{22}=\frac{e^{-j\pi}}{2}(S_{22a}-S_{22b})=0 \quad ,(VSWR)_{OUT}=1$$

$$G_T=(0.5)^2|S_{21a}+S_{21b}|^2=(0.5)^2[2(3.4)]^2=11.56 \text{ OR } 10.6\text{ dB}$$

(b) $S_{11b}=0.525\angle 168^\circ,\ S_{12b}=0.084\angle 63^\circ,\ S_{21b}=3.57\angle 73.5^\circ,\ S_{22b}=0.42\angle -42.75^\circ$

$$S_{11}=\frac{e^{-j\pi}}{2}(0.5\angle 160^\circ-0.525\angle 168^\circ)=0.038\angle -125.2^\circ \quad ,(VSWR)_{in}=\frac{1+0.038}{1-0.038}=1.08$$

$$S_{22}=\frac{e^{-j\pi}}{2}(0.4\angle -45^\circ-0.42\angle -42.75^\circ)=0.013\angle -5^\circ \quad ,(VSWR)_{OUT}=\frac{1+0.013}{1-0.013}=1.03$$

$$G_T=(0.5)^2|3.4\angle 70^\circ+3.57\angle 73.5^\circ|^2=12.13 \text{ OR } 10.8\text{ dB}$$

4.17) (a) FOR IDENTICAL AMPLIFIERS: $S_{11}=S_{22}=0$. THEN,

$$K=\frac{1+|\Delta|^2}{2|S_{12}S_{21}|}=\frac{1+|S_{12}S_{21}|^2}{2|S_{12}S_{21}|}=\frac{1+P^2}{2P} \quad , P=|S_{12}S_{21}|$$

THE MINIMUM VALUE OF K OCCURS WHEN $P=1$, AND IT IS $K=1$.

(b) $S_{11}=0$, $S_{22}=0$

$$S_{21}=\frac{e^{-j\frac{\pi}{2}}}{2}(S_{21a}+S_{21b})=\frac{e^{-j\frac{\pi}{2}}}{2}2(-10.9+j7.895)=7.895+j10.9$$

$$S_{12}=\frac{e^{-j\frac{\pi}{2}}}{2}(S_{12a}+S_{12b})=\frac{e^{-j\frac{\pi}{2}}}{2}2(0.009+j0.015)=0.015-j0.009$$

$$P=|S_{12}S_{21}|=|0.017(13.46)|=0.235 \quad , |\Delta|<1 \text{ AND}$$

$$K=\frac{1+(0.235)^2}{2(0.235)}=2.25 \quad \therefore \text{UNCONDITIONALLY STABLE.}$$

4.18)
$$\begin{bmatrix} i_1 \\ i_2 \end{bmatrix}=\begin{bmatrix} 1/R_2 & -1/R_2 \\ \frac{g_m}{1+g_mR_1}-\frac{1}{R_2} & \frac{1}{R_2} \end{bmatrix}\begin{bmatrix} v_1 \\ v_2 \end{bmatrix}$$

FROM TABLE 1.8.1: $y'_{11}=y_{11}Z_0=\frac{Z_0}{R_2}$, $y'_{12}=y_{12}Z_0=-\frac{Z_0}{R_2}$,

$y'_{21}=y_{21}Z_0=\frac{g_mZ_o}{1+g_mR_1}-\frac{Z_0}{R_2}$, $y'_{22}=y_{22}Z_0=\frac{Z_0}{R_2}$

$$\Delta_3=(1+y'_{11})(1+y'_{22})-y'_{12}y'_{21}=\left(1+\frac{Z_0}{R_2}\right)^2+\frac{Z_0}{R_2}\left(\frac{g_mZ_o}{1+g_mR_1}-\frac{Z_0}{R_2}\right)$$

OR

$$\Delta_3=1+\frac{2Z_0}{R_2}+\frac{g_mZ_o^2}{R_2(1+g_mR_1)}$$

$$S_{11} = \frac{(1-y'_{11})(1+y'_{22}) + y'_{12}y'_{12}}{\Delta_3} = \frac{1}{\Delta_3}\left[\left(1-\frac{Z_0}{R_2}\right)\left(1+\frac{Z_0}{R_2}\right) + \frac{Z_0}{R_2}\left(\frac{Z_0}{R_2} - \frac{g_m Z_0}{1+g_m R_1}\right)\right]$$

$$\therefore \quad S_{11} = \frac{1}{\Delta_3}\left[1 - \frac{g_m Z_0^2}{R_2(1+g_m R_1)}\right]$$

$$S_{22} = \frac{(1+y'_{11})(1-y'_{22}) + y'_{12}y'_{21}}{\Delta_3} = \frac{1}{\Delta_3}\left[1 - \frac{g_m Z_0^2}{R_2(1+g_m R_1)}\right]$$

$$S_{21} = \frac{-2y'_{21}}{\Delta_3} = \frac{-2}{\Delta_3}\left[\frac{g_m Z_0}{1+g_m R_1} - \frac{Z_0}{R_2}\right] = \frac{1}{\Delta_3}\left[\frac{-2g_m Z_0}{1+g_m R_1} + \frac{2Z_0}{R_2}\right]$$

$$S_{12} = \frac{-2y'_{12}}{\Delta_3} = \frac{-2}{\Delta_3}\left(-\frac{Z_0}{R_2}\right) = \frac{2Z_0}{\Delta_3 R_2}$$

4.19) IF S_{11} IS APPROXIMATED BY $0.97\angle 0°$, THE ASSOCIATED $r_{b'e}$ IS 3.28 kΩ.

$$G_T = 10 = |S_{21}|^2 \Rightarrow S_{21} = -3.16$$

FROM (4.4.10):

$$R_2 = Z_0(1-S_{21}) = 50(1+3.16) = 208\ \Omega$$

THE VALUE OF g_m FOR $(VSWR)_{in} \approx 1$ AND $(VSWR)_{OUT} \approx 1$ FOLLOWS FROM (4.4.9). THAT IS,

$$g_m = \frac{R_2}{Z_0^2} = \frac{208}{50^2} = 83\ mS$$

4.20) $G_T = 10 = |S_{21}|^2 \Rightarrow S_{21} = -3.16$.

FROM (4.4.10): $R_2 = Z_0(1-S_{21}) = 50(1+3.16) = 208\ \Omega$

AT THE LOWER FREQUENCIES THE VALUE OF $S_{11} = 0.651\angle -74°$ $(6+j14\ mS)$ CORRESPONDS TO AN $r_{b'e}$ OF 166.6 Ω. HENCE, THE CONDITION $r_{b'e} \gg R_2$ IS NOT SATISFIED, AND A SERIES FEEDBACK RESISTOR R_1 IS NEEDED.

FROM (4.4.10): $g_{m(min)} = \frac{1+3.16}{50} = 83\ mS$

THE VALUE OF THE TRANSISTOR g_m AT THE LOWER FREQUENCIES (WITH $S_{21} \approx 34\angle 180°$) IS:

$$g_m = -\frac{S_{21}}{2Z_0} = \frac{34}{2(50)} = 340\ mS$$

SINCE $g_m > g_{m(min)}$, R_1 IS CALCULATED USING (4.4.7):

$$R_1 = \frac{Z_0^2}{R_2} - \frac{1}{g_m} = \frac{50^2}{208} - \frac{1}{0.34} = 9\ \Omega$$

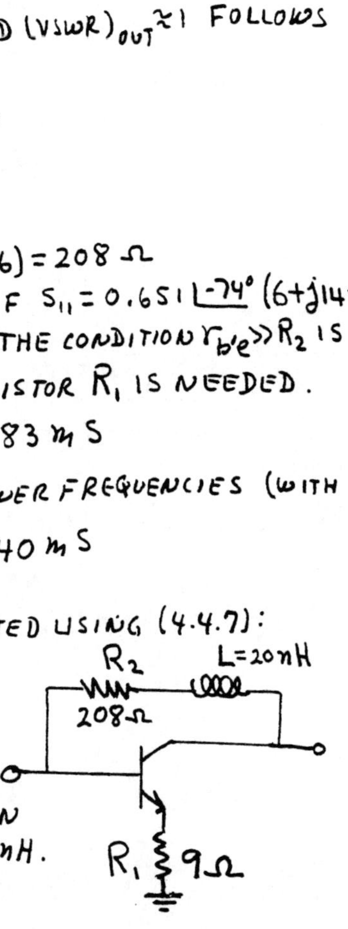

AN INDUCTOR IS USUALLY NEEDED IN SERIES WITH R_2 IN ORDER TO REDUCE THE FEEDBACK AT THE HIGHER FREQUENCIES AND TO FLATTEN THE GAIN. TYPICAL VALUES OF L ARE 20nH TO 30nH.

(b) THE FEEDBACK CIRCUIT WAS SIMULATED USING THE HEWLETT-PACKARD MDS PROGRAM. THE RESULTS ARE:

f(MHz)	S_{11}	S_{21}	S_{12}	S_{22}	$G_T = 10\log\lvert S_{21}\rvert^2$
100	$0.057\angle{-163^\circ}$	$3.27\angle{177.3^\circ}$	$0.183\angle{-1.86^\circ}$	$0.044\angle{113.4^\circ}$	10.3 dB
200	$0.062\angle{-146.6^\circ}$	$3.29\angle{174.5^\circ}$	$0.182\angle{-3.8^\circ}$	$0.083\angle{95.4^\circ}$	10.3 dB
400	$0.086\angle{-120.3^\circ}$	$3.35\angle{168.6^\circ}$	$0.177\angle{-7.6^\circ}$	$0.161\angle{78.4^\circ}$	10.5 dB
600	$0.132\angle{-105.3^\circ}$	$3.45\angle{161.7^\circ}$	$0.169\angle{-11.4^\circ}$	$0.243\angle{66.4^\circ}$	10.8 dB
800	$0.196\angle{-102.9^\circ}$	$3.52\angle{154^\circ}$	$0.157\angle{-14.4^\circ}$	$0.32\angle{54.1^\circ}$	10.9 dB
1000	$0.27\angle{-104.5^\circ}$	$3.56\angle{145.3^\circ}$	$0.142\angle{-16.8^\circ}$	$0.393\angle{42.4^\circ}$	11.0 dB
1500	$0.46\angle{-117.7^\circ}$	$3.38\angle{121.9^\circ}$	$0.102\angle{-11.6^\circ}$	$0.518\angle{13.6}$	10.6 dB

4.21) (a)

$$Q_1 = \frac{X_c}{R} = \frac{1}{\omega_0 C R} = \frac{1}{2\pi\, 500\,10^6 (100\,10^{-2})5} = 0.637$$

$$Q_2 = \frac{f_o}{f_2 - f_1} = \frac{500\,10^6}{600\,10^6 - 400\,10^6} = 2.5$$

$$\Gamma_x = e^{-\pi(Q_2/Q_1)} = e^{-\pi(2.5/0.637)} = 4.4\ 10^{-6}$$

(b)

$$Q_1 = \frac{R}{X_c} = \omega_0 RC = 2\pi\,(9\,10^9)\,50\,(10^{-12}) = 2.83$$

$$Q_2 = \frac{f_o}{f_2 - f_1} = \frac{9\,10^9}{12\,10^9 - 6\,10^9} = 1.5$$

$$\Gamma_x = e^{-\pi(Q_2/Q_1)} = e^{-\pi(1.5/2.83)} = 0.19$$

4.22)

$$\int_0^\infty \frac{1}{\omega^2} \ln\left|\frac{1}{\Gamma}\right| d\omega \le \pi RC \Rightarrow \ln\left|\frac{1}{\Gamma_x}\right| \int_{\omega_a}^{\omega_b} \frac{d\omega}{\omega^2} = \pi RC$$

$$\text{OR } \ln|\Gamma_x|\left(\frac{1}{\omega_b} - \frac{1}{\omega_a}\right) = \pi RC \Rightarrow \ln|\Gamma_x| = \frac{\omega_a \omega_b \pi RC}{\omega_a - \omega_b} = \frac{-\pi \omega_0^2 RC}{\omega_b - \omega_a}$$

$$\text{OR } \ln|\Gamma_x| = -\pi \frac{Q_2}{Q_1} \quad \text{WHERE } \omega_0 = \sqrt{\omega_a \omega_b},\ Q_2 = \frac{\omega_0}{\omega_b - \omega_a},\ Q_1 = \frac{1}{\omega_0 RC}$$

$\therefore |\Gamma_x| = e^{-\pi Q_2/Q_1}$, ω_0 IS THE GEOMETRIC MEAN OF ω_a AND ω_b.

FOR: $\int_0^\infty \frac{1}{\omega^2} \ln\left|\frac{1}{\Gamma}\right| d\omega \le \frac{\pi L}{R} \Rightarrow \ln|\Gamma_x|\left(\frac{1}{\omega_b} - \frac{1}{\omega_a}\right) = \frac{\pi L}{R}$

OR $\ln|\Gamma_x| = \frac{-\pi \omega_0^2 L}{(\omega_b - \omega_a) R} = -\pi \frac{Q_2}{Q_1} \Rightarrow |\Gamma_x| = e^{-\pi Q_2/Q_1}$

WHERE $\omega_0 = \sqrt{\omega_a \omega_b}$, $Q_2 = \frac{\omega_0}{\omega_b - \omega_a}$, $Q_1 = \frac{R}{\omega_0 L}$

FOR: $\int_0^\infty \ln\left|\frac{1}{\Gamma}\right| d\omega \le \frac{\pi R}{L} \Rightarrow \ln|\Gamma_x| = \frac{-\pi R}{(\omega_b - \omega_a)} = -\pi \frac{Q_2}{Q_1} \Rightarrow |\Gamma_x| = e^{-\pi Q_2/Q_1}$

WHERE $Q_2 = \frac{\omega_0}{\omega_b - \omega_a}$, $Q_1 = \frac{\omega_0 L}{R}$

4.23) (a) $\Gamma_{IN} = S_{11} + \frac{S_{12} S_{21} \Gamma_L}{1 - S_{22}\Gamma_L} = \frac{S_{11} - \Delta \Gamma_L}{1 - S_{22}\Gamma_L}$

$$\frac{\partial \Gamma_{IN}}{\partial \Gamma_L} = \frac{(1 - S_{22}\Gamma_L) S_{12} S_{21} + S_{12} S_{21} \Gamma_L S_{22}}{(1 - S_{22}\Gamma_L)^2}$$

$$\delta_{IN} = \left|\frac{d\Gamma_{IN}/\Gamma_{IN}}{d\Gamma_L/\Gamma_L}\right| = \left|\frac{\Gamma_L (1 - S_{22}\Gamma_L)\left[(1 - S_{22}\Gamma_L) S_{12} S_{21} + S_{12} S_{21} S_{22} \Gamma_L\right]}{(S_{11} - \Delta\Gamma_L)\ (1 - S_{22}\Gamma_L)^2}\right|$$

$$\delta_{IN} = \left|\frac{\Gamma_L (1 - S_{22}\Gamma_L) S_{12} S_{21} + S_{12} S_{21} S_{22} \Gamma_L^2}{(S_{11} - \Delta\Gamma_L)(1 - S_{22}\Gamma_L)}\right| = \frac{|S_{21}||S_{12}||\Gamma_L|}{|1 - S_{22}\Gamma_L||S_{11} - \Delta\Gamma_L|}$$

(b) $|1 - S_{22}\Gamma_L||S_{11} - \Delta\Gamma_L| = |S_{21} S_{12}| \delta_{IN}^{-1} |\Gamma_L|$

$|S_{11} - \Delta\Gamma_L - S_{11} S_{22}\Gamma_L + S_{22}\Delta\Gamma_L^2 - S_{21} S_{12} \delta_{IN}^{-1} \Gamma_L| = 0$

$\left|\Gamma_L^2 - \Gamma_L\left(\frac{\Delta + S_{11} S_{22} + S_{12} S_{21} \delta_{IN}^{-1}}{S_{22}\Delta}\right) + \frac{S_{11}}{S_{22}\Delta}\right| = 0$

$\therefore |\Gamma_L| = \left|\alpha \pm \left|\alpha^2 - \frac{S_{11}}{S_{22}\Delta}\right|^{1/2}\right|$ WHERE $\alpha = \frac{\Delta + S_{11} S_{22} + S_{12} S_{21} \delta_{IN}^{-1}}{2 S_{22} \Delta}$

(c) $\Gamma_{OUT} = S_{22} + \frac{S_{12} S_{21} \Gamma_s}{1 - S_{11}\Gamma_s} = \frac{S_{22} - \Delta\Gamma_s}{1 - S_{11}\Gamma_s}$

$$\frac{\partial \Gamma_{OUT}}{\partial \Gamma_s} = \frac{(1 - S_{11}\Gamma_s) S_{12} S_{21} + S_{12} S_{21} \Gamma_s S_{11}}{(1 - S_{11}\Gamma_s)^2}$$

$$\delta_{OUT} = \left|\frac{\partial\Gamma_{OUT}/\Gamma_{OUT}}{\partial\Gamma_s/\Gamma_s}\right| = \left|\frac{\Gamma_s (1 - S_{11}\Gamma_s)\left[(1 - S_{11}\Gamma_s) S_{12} S_{21} + S_{12} S_{21} S_{11} \Gamma_s\right]}{(S_{22} - \Delta\Gamma_s)\ (1 - S_{11}\Gamma_s)^2}\right|$$

$$\delta_{OUT} = \left|\frac{\Gamma_s (1 - S_{11}\Gamma_s) S_{12} S_{21} + S_{12} S_{21} S_{11} \Gamma_s^2}{(S_{22} - \Delta\Gamma_s)(1 - S_{11}\Gamma_s)}\right| = \frac{|S_{21}||S_{12}||\Gamma_s|}{|1 - S_{11}\Gamma_s||S_{11} - \Delta\Gamma_s|}$$

4.24) (a) $\Gamma_s = S_{11}^* = 0.75\angle 100^\circ$ AND $\Gamma_L = S_{22}^* = 0.7\angle 50^\circ$

(b) $Y_s = \frac{1}{Z_s} = 7 - j23$ mS , $Y_L = \frac{1}{Z_L} = 4 - j9$ mS

$$(BW)_{IN}^{i} = \frac{2f_o G_{s,m}}{|B_{s,m}|} = \frac{2(8\,10^9)\,0.007}{0.023} = 4.87 \text{ GHz}$$

$$(BW)_{OUT}^{i} = \frac{2f_o G_{L,m}}{|B_{L,m}|} = \frac{2(8\,10^9)(0.004)}{0.009} = 7.11 \text{ GHz}$$

(c) $(BW)_{IN} = 20\%\,(BW)_{IN}^{i} = 974$ MHz

$Y_{IN} = 7 + j23$ mS

THE REQUIRED VALUE OF C_{IN}' (FROM (4.6.5)) IS:

$$C_{IN}' = \frac{B_{IN,m}}{\omega_0}\left[\frac{(BW)_{IN}^{i}}{(BW)_{IN}} - 1\right] = \frac{23\,10^{-3}}{2\pi(8\,10^9)}\left[\frac{4.87}{0.974} - 1\right] = 1.83 \text{ pF}$$

(d) $Y_{IN}' = Y_{IN} + j\omega C_{IN}' = (7 + j23)10^{-3} + j2\pi(8\,10^9)(1.83\,10^{-12}) = 7 + j115$ mS

$\Gamma_{IN}' = 0.98\angle -160.4^\circ$, $\therefore \Gamma_s = \Gamma_{IN}'^{*} = 0.98\angle 160.4^\circ$, $\Gamma_L = S_{22}^* = 0.7\angle 50^\circ$

$$G_{TU,max} = \frac{1}{1-(0.98)^2}(2.5)^2\frac{1}{1-(0.7)^2} = 309.4 \text{ OR } 24.9 \text{ dB}$$

4.25) $\Gamma_{Ms} = 0.476\angle 166^\circ \Rightarrow Y_{Ms} = 51 - j15$ mS

$\Gamma_{ML} = 0.846\angle 72^\circ \Rightarrow Y_{ML} = 3 - j14$ mS

$$(BW)_{IN}^{i} = \frac{2(4\,10^9)\,0.051}{0.015} = 27.2 \text{ GHz} \quad , \quad (BW)_{OUT}^{i} = \frac{2(4\,10^9)\,0.003}{0.014} = 1.714 \text{ GHz}$$

THE VALUE OF L_{OUT}' REQUIRED FOR $BW \approx (BW)_{OUT} = 400$ MHz IS

$$L_{OUT}' = \frac{1}{\omega_0 |B_{OUT,m}|\left[\frac{(BW)_{OUT}^{i}}{(BW)_{OUT}} - 1\right]} = \frac{1}{2\pi(4\,10^9)0.014\left[\frac{1.714\,10^9}{0.4\,10^9} - 1\right]} = 0.865 \text{ nH}$$

4.26) FROM (4.7.4): $P_{i,mds} = -174 + 10\log 80\,10^6 + 5 + 3 = -76.97$ dBm

FROM (4.7.5): $P_{o,mds} = P_{i,mds} + G_A = -76.97 + 30 = -46.97$ dBm.
(WITH $G_A = G_T$)

$$DR = P_{1dB} - P_{o,mds} = 28 - (-46.97) = 74.97 \text{ dB}$$

$$P_{IP} = P_{1dB} + 10 = 28 + 10 = 38 \text{ dBm}$$

$$DR_f = \frac{2}{3}(P_{IP} - P_{o,mds}) = \frac{2}{3}(38 - (-46.97)) = 69.3 \text{ dB}$$

FOR NO THIRD-ORDER IM:

$$P_{OUT} = P_{o,mds} + DR_f = -46.97 + 69.3 = 22.3 \text{ dBm}$$

4.27)

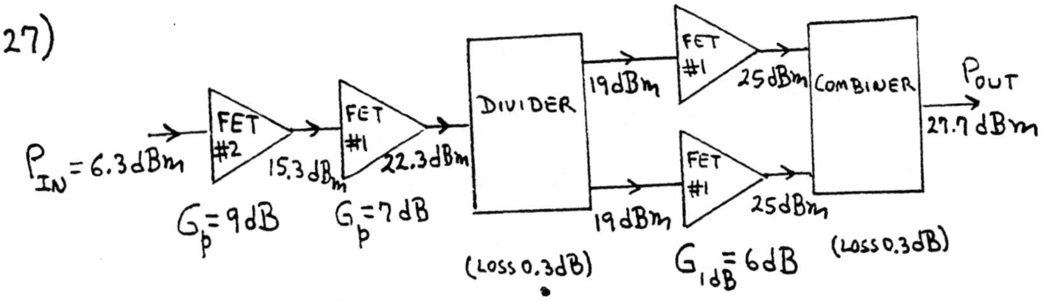

4.28) $K = 1.61$, $\Delta = 0.313\angle 18.8^\circ$ ∴ UNCONDITIONALLY STABLE

$\Gamma_L = \Gamma_{Lp} = 0.1\angle 0^\circ$ AND $\Gamma_s = \Gamma_{sp} = \Gamma_{IN}^* = 0.324\angle 146.6^\circ$

OUTPUT MATCHING CIRCUIT:

$Z_{Lp} = 61.1\ \Omega$

USING A $\lambda/4$ TRANSFORMER

$Z_0 = \sqrt{61.1(50)} = 55.3\ \Omega$

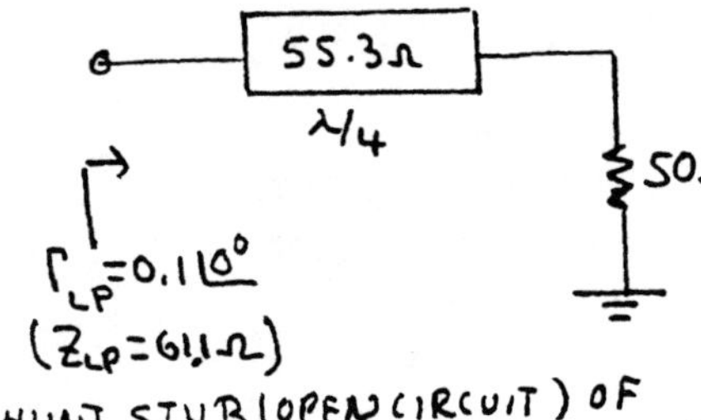

INPUT MATCHING CIRCUIT: USE A SHUNT STUB (OPEN CIRCUIT) OF LENGTH 0.097λ FOLLOWED BY A SERIES TRANS. LINE OF LENGTH 0.144λ.

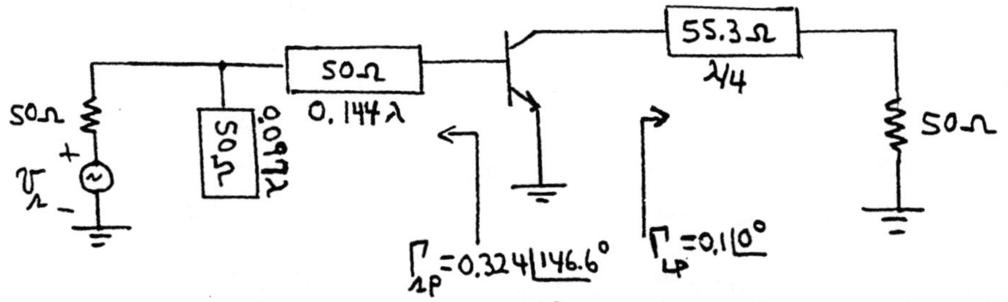

4.29) $P_{IN} = \dfrac{E_{s,rms}^2}{4(50)} = \dfrac{(79.5\,10^{-3})^2}{200} = 3.16\,10^{-5}$ W OR -15 dBm

$P_{OUT} = P_{IN} + 13 + 13 + 10 + 9 = -15 + 45 = 30$ dBm

OR $P_{OUT} = 1$ W

4.30) THE BLOCK DIAGRAM IS IDENTICAL TO THE ONE IN FIG. 4.7.18, EXCEPT THAT $Z_s = 1.19 + j1.35\ \Omega$ AND $Z_L = 8.1 - j4.1\ \Omega$

4.31) YES.

5.1) $\beta(j\omega)A_{vo}=1$ OR $[\beta_r(j\omega)+j\beta_i(j\omega)]A_{vo}=1$

$\therefore A_{vo}=\dfrac{1}{\beta_r(\omega)}$ AND $\beta_i(j\omega)=0$

$$\beta(j\omega)=\frac{10^{-5}}{1+(\omega-10^3)^2}-j\frac{10^{-5}(\omega-10^3)}{1+(\omega-10^3)^2}$$

$\beta_i(j\omega)=0$ WHEN $\omega=10^3$ rad/s

AT $\omega=\omega_0=10^3$ rad/s : $\beta_r(j\omega_0)=10^{-5}$

THEN, $A_{vo}=\dfrac{1}{10^{-5}}=10^5$

5.2) $X_L(j\omega_0)=j\omega_0 L+\dfrac{1}{j\omega_0 C}=0$ OR $\omega_0=\dfrac{1}{\sqrt{LC}}$

$$f_o=\frac{1}{2\pi\sqrt{50\,10^{-9}\,10\,10^{-12}}}=225\text{ MHz}$$

$$R_L=\frac{R_o}{3}=\frac{30}{3}=10\,\Omega$$

5.3) $B_L(j\omega_0)=j\omega_0 C+\dfrac{1}{j\omega_0 L}=0$ OR $\omega_0=\dfrac{1}{\sqrt{LC}}$

$$f_o=\frac{1}{2\pi\sqrt{25\,10^{-9}\,5\,10^{-12}}}=450\text{ MHz}$$

$$G_L=\frac{G_o}{3}=\frac{40\,10^{-3}}{3}=13.3\text{ mS}$$

5.4) (a) $Z_{IN}(A',\omega)=\dfrac{1}{G(A')+j\omega C}=\dfrac{G(A')}{G^2(A')+\omega^2C^2}+j\dfrac{-\omega C}{G^2(A')+\omega^2C^2}$

$\therefore R_{IN}(A',\omega)=\dfrac{G(A')}{G^2(A')+\omega^2C^2}$ AND $X_{IN}(A',\omega)=\dfrac{-\omega C}{G^2(A')+\omega^2C^2}$

FROM (5.2.16), (5.2.17), AND (5.2.20), A STABLE OSCILLATION OCCURS

$$R_L=-R_{IN}(A',\omega)=\frac{-G(A')}{G^2(A')+\omega^2C^2}$$

$$X_L=-X_{IN}(A',\omega)=\frac{\omega C}{G^2(A')+\omega^2C^2}$$

AND

$$\left.\frac{\partial R_{IN}}{\partial A'}\right|_{A'=A_o}\left.\frac{\partial X_L}{\partial \omega}\right|_{\omega=\omega_0}>0$$

(b) $P=\frac{1}{2}|V|^2|G(A')|=\frac{1}{2}A'^2G_o(1-A'/A'_M)$, $\dfrac{\partial P}{\partial A'}=\dfrac{G_o}{2}\left(2A'-\dfrac{3A'^2}{A'_M}\right)=0$

OR $A'\equiv A'_{o,max}=\frac{2}{3}A'_M$. AT $A'=A'_{o,max}$ THE VALUE OF $G_{IN}(A')$ IS:

$G_{IN}(A_{o,max})=-G_o/3$. THEREFORE: $G_L=G_o/3$.

5.5) $K = -0.505$, $\Delta = 0.933\angle 141.9^\circ$ ∴ POTENTIALLY UNSTABLE

INPUT STABILITY CIRCLE: $C_\lambda = 1.56\angle 2.55^\circ$, $r_\lambda = 0.789$

OUTPUT STABILITY CIRCLE: $C_L = 0.89\angle -155.2^\circ$, $r_L = 0.295$

SELECT THE GATE TO SOURCE AS THE TERMINATING PORT. THE VALUE FOR Γ_T SELECTED IS $\Gamma_T = 0.5\angle -145^\circ$.

THEN, $\Gamma_{IN} = 1.116\angle 173^\circ$

$$Z_{IN} = -2.74 + j3.03\ \Omega$$

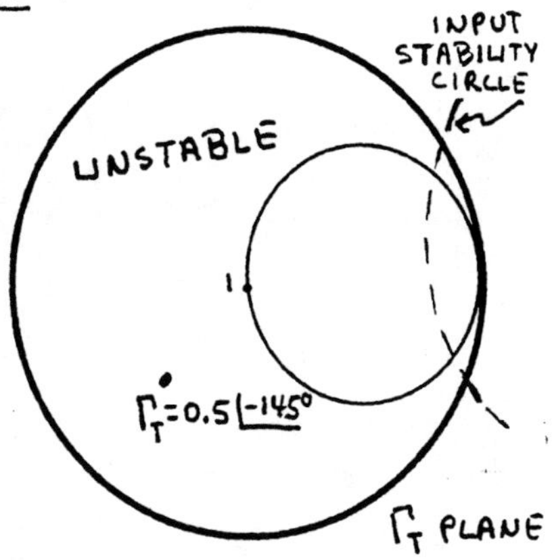

LET $Z_L = \frac{2.74}{3} - j3.03 = 0.91 - j3.03\,\Omega$

NEGLECTING THE $0.91\,\Omega$, Z_L CAN BE IMPLEMENTED WITH AN OPEN-CIRCUITED LINE OF LENGTH 0.24λ.

A DESIGN FOR THE OSCILLATOR IS:

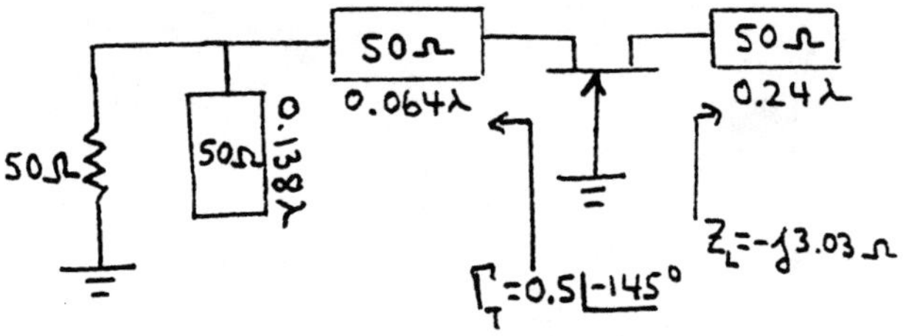

5.6) $Z_L = 19 + j2.6\ \Omega$, $z_L = 0.38 + j0.052$, $\Gamma_L = 0.451\angle 173^\circ$

$$y_L = \frac{1}{z_L} = 2.58 - j0.353$$

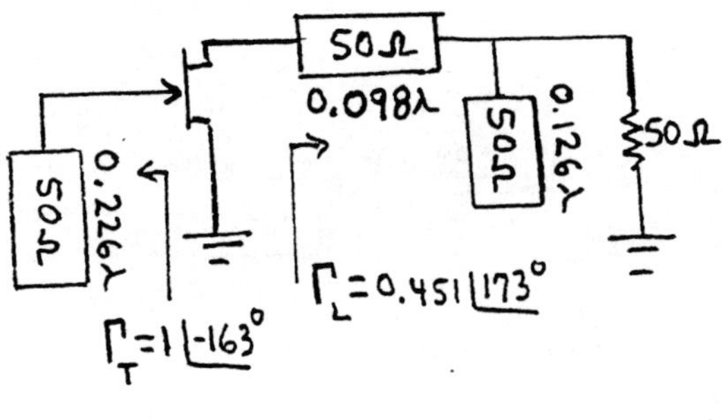

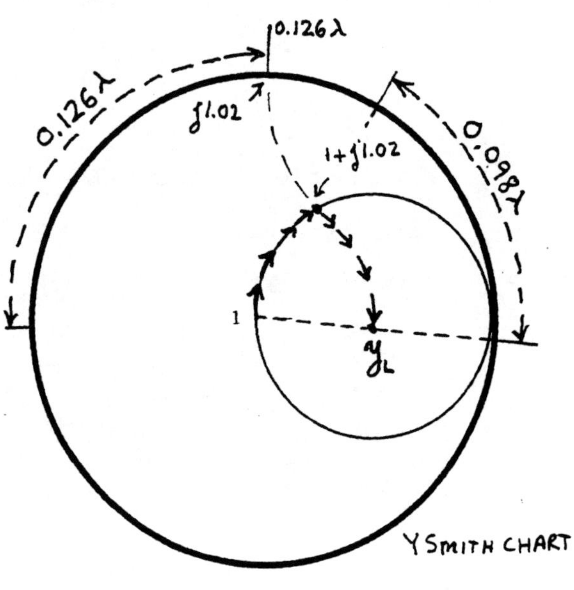

5.7) WITH $L = 0.5\,nH$: $K = -0.834$, $\Delta = 1.02\angle 157°$ ∴ POTENTIALLY UNSTABLE

INPUT STABILITY CIRCLE : $C_s = 1.38\angle -50.9°$, $r_s = 2.1$

OUTPUT STABILITY CIRCLE : $C_L = 0.78\angle 119.7°$, $r_L = 1.39$

SELECTING THE OUTPUT PORT (i.e., BASE TO COLLECTOR) AS THE TERMINATING PORT WITH:

$\Gamma_T = 0.5\angle 30°$

THEN: $\Gamma_{IN} = 1.126\angle 172.4°$

$Z_{IN} = -2.98 + j3.31\ \Omega$

LET $Z_L = \frac{2.98}{3} - j3.31 = 0.99 - j3.31\ \Omega$

NEGLECTING THE 0.99Ω WE CAN DESIGN THE LOAD CIRCUIT WITH A CAPACITOR, NAMELY

$-j3.31 = \frac{-j}{\omega C}$ OR $C = \frac{1}{2\pi(2\times10^9)3.31} = 24\,pF$

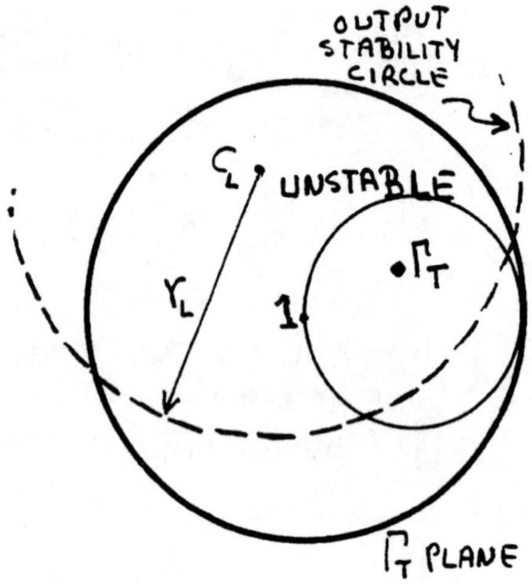

A DESIGN FOR THE OSCILLATOR IS:

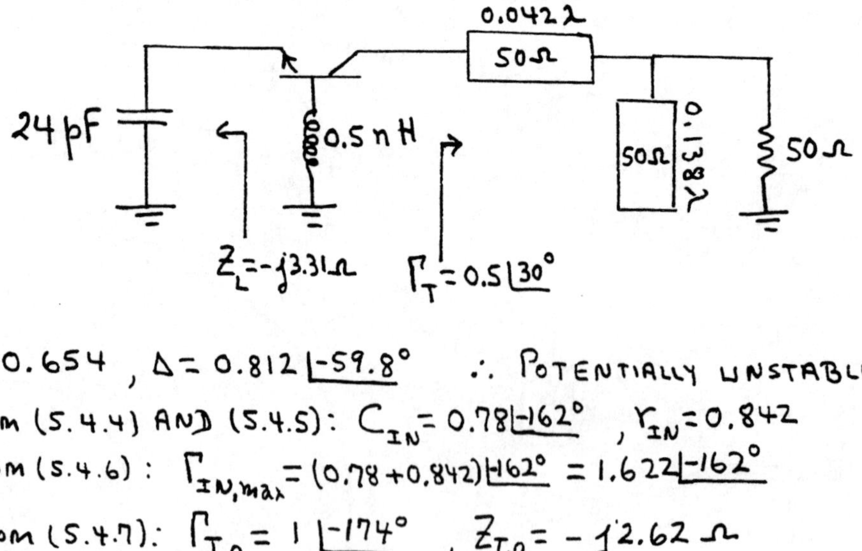

5.8) $K = 0.654$, $\Delta = 0.812\angle -59.8°$ ∴ POTENTIALLY UNSTABLE

FROM (5.4.4) AND (5.4.5): $C_{IN} = 0.78\angle -162°$, $r_{IN} = 0.842$

FROM (5.4.6): $\Gamma_{IN,max} = (0.78 + 0.842)\angle -162° = 1.622\angle -162°$

FROM (5.4.7): $\Gamma_{T,o} = 1\angle -174°$, $Z_{T,o} = -j2.62\ \Omega$

$Z_{IN,max} = -12.1 - j7.5\Omega$

THE LARGE-SIGNAL CHARACTERISTICS SHOW THAT A POWER OF 150 mW IS POSSIBLE WITH $Z_L = 50(0.2 - j0.35) = 10 - j17.5\ \Omega$

THE BLOCK DIAGRAM OF THE OSCILLATOR IS:

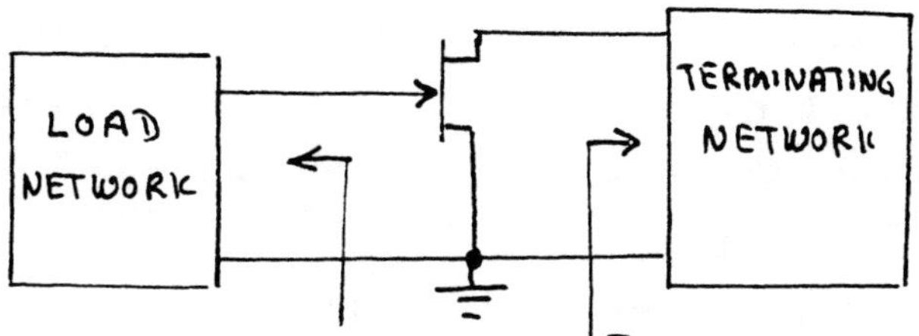

5.9) $$\Gamma_{IN} = \frac{S_{11} - \Delta\Gamma_T}{1 - S_{22}\Gamma_T} = \frac{S_{11}(1-|S_{22}|^2) - \Delta\Gamma_T(1-|S_{22}|^2)}{(1-S_{22}\Gamma_T)(1-|S_{22}|^2)}$$

$$\Gamma_{IN} = \frac{S_{11} - S_{11}S_{22}S_{22}^* - S_{11}S_{22}\Gamma_T + S_{21}S_{12}\Gamma_T + \Delta\Gamma_T\, S_{22}S_{22}^*}{(1-S_{22}\Gamma_T)(1-|S_{22}|^2)}$$

ADDING AND SUBTRACTING IN THE NUMERATOR THE TERM $S_{21}S_{12}S_{22}^*$ GIVES

$$\Gamma_{IN} = \frac{S_{11} - \Delta S_{22}^* - S_{11}S_{22}\Gamma_T + \Delta\Gamma_T S_{22}S_{22}^* + S_{21}S_{12}(\Gamma_T - S_{22}^*)}{(1-S_{22}\Gamma_T)(1-|S_{22}|^2}$$

$$\Gamma_{IN} = \frac{S_{11}(1-S_{22}\Gamma_T) - \Delta S_{22}^*(1-S_{22}\Gamma_T) + S_{21}S_{12}(\Gamma_T - S_{22}^*)}{(1-S_{22}\Gamma_T)(1-|S_{22}|^2)}$$

$$\Gamma_{IN} = \frac{S_{11} - \Delta S_{22}^*}{1-|S_{22}|^2} + \frac{S_{12}S_{21}}{1-|S_{22}|^2}\;\frac{\Gamma_T - S_{22}^*}{1-S_{22}\Gamma_T} = \Gamma_{IN,0} + \alpha\Gamma_T'$$

WHERE $\Gamma_{IN,0} = \dfrac{S_{11} - \Delta S_{22}^*}{1-|S_{22}|^2}$ AND $\alpha\Gamma_T' = \dfrac{S_{12}S_{21}}{1-|S_{22}|^2}\;\dfrac{\Gamma_T - S_{22}^*}{1-S_{22}\Gamma_T}$

IF α IS DEFINED AS: $\alpha = \dfrac{S_{12}S_{21}}{1-|S_{22}|^2}\;\dfrac{1-S_{22}^*}{1-S_{22}}$, THEN

$\Gamma_T' = \dfrac{\Gamma_T - S_{22}^*}{1-S_{22}\Gamma_T}\;\dfrac{1-S_{22}}{1-S_{22}^*}$. LETTING $\Gamma_T = \dfrac{Z_T - Z_0}{Z_T + Z_0}$ AND $S_{22} = \dfrac{Z_{22} - Z_0}{Z_{22} + Z_0}$

$$\therefore\ \Gamma_T' = \frac{\left(\frac{Z_T - Z_0}{Z_T + Z_0}\right) - \left(\frac{Z_{22}^* - Z_0}{Z_{22}^* + Z_0}\right)}{1 - \left(\frac{Z_{22} - Z_0}{Z_{22} + Z_0}\right)\left(\frac{Z_T - Z_0}{Z_T + Z_0}\right)}\left(\frac{Z_{22}^* + Z_0}{Z_{22} + Z_0}\right) = \frac{Z_T - Z_{22}^*}{Z_T + Z_{22}}$$

$\Gamma_{IN,max}$ (SEE FIG. 5.4.3) OCCURS WHEN $|\Gamma_T'| = 1$ AND $\angle\alpha = \angle\Gamma_{IN,0}$. HENCE,

$$\Gamma_{IN,max} = \Gamma_{IN,0} + |\alpha|\,\hat{u}_{IN,o} \qquad (1)$$

WHERE $\hat{u}_{IN,o}$ IS A UNIT VECTOR IN THE DIRECTION OF $\Gamma_{IN,0}$.

(b) (1) IS IDENTICAL TO (5.4.6) BECAUSE $\Gamma_{IN,0} = C_{IN}$, $|\alpha| = |r_{IN}|$, AND $\hat{u}_{IN,o} = \angle C_{IN}$. (1) CAN ALSO BE WRITTEN AS

$$\Gamma_{IN,max} = (|\Gamma_{IN,0}| + |\alpha|)\angle\Gamma_{IN,0}$$

WHICH IS THE SAME AS (5.4.6).

5.10) THE S PARAMETERS OF A SERIES IMPEDANCE Z IN A Z_0 SYSTEM WERE DERIVED IN EXAMPLE 1.6.1. IN THE CASE OF A DRO THE VALUE OF Z IS GIVEN BY 5.5.6, NAMELY

$$Z = \frac{R}{1+j2Q_u\delta} = \frac{2\beta Z_0}{1+j2Q_u\delta}$$

USING (1.6.24) AND (1.6.25) WE OBTAIN:

$$S_{11} = S_{22} = \frac{Z}{Z+2Z_0} = \frac{\beta}{\beta+1+j2Q_u\delta}$$

$$S_{21} = S_{12} = \frac{2Z_0}{Z+2Z_0} = \frac{1+j2Q_u\delta}{\beta+1+j2Q_u\delta}$$

5.11) AT 12 GHz: $K = 1.36$ AND $\Delta = 0.41\angle 139.8^\circ$, ∴ UNCONDITIONALLY STABLE.

THE S PARAMETERS ARE CLOSE TO THOSE USED IN EXAMPLE 5.5.1. HENCE, THE DESIGN PROCEDURE IS SIMILAR TO THE ONE USED IN EXAMPLE 5.5.1.

USING A SERIES FEEDBACK CAPACITOR (SEE FIG. 5.5.12a) WITH $Z = -j120\,\Omega$ RESULTS IN A POTENTIALLY UNSTABLE CONFIGURATION WITH: $S_{11} = 1.255\angle 117^\circ$, $S_{12} = 1.65\angle -78.4^\circ$, $S_{21} = 1.87\angle -23.2^\circ$, AND $S_{22} = 0.9\angle 102^\circ$.

THE GATE TO GROUND PORT IS SELECTED AS THE TERMINATING PORT. THE STABILITY CIRCLE AT THE TERMINATING PORT IS SIMILAR TO THE ONE DRAWN IN FIG. 5.5.12b (i.e., $C_\Omega = 0.773\angle 26.5^\circ$ AND $r_\Omega = 0.805$).

WITH $\beta = 10$: $R = 10(2\times 50) = 1000\,\Omega$ AND $\Gamma_T = \frac{10}{10+1}e^{-j2\theta} = \frac{10}{11}e^{-j2\theta}$

LETTING $l_1 = \lambda/4$ (i.e. $\theta = \pi/2$), THEN $\Gamma_T = 0.909\angle -180^\circ$, $\Gamma_{IN} = 2.68\angle 33.4^\circ$ (OR $Z_{IN} = -83.4 + j39.8\,\Omega$), AND Z_L IS SELECTED AS: $Z_L = 27.8 - j39.8\,\Omega$

THE DRO CIRCUIT IS SHOWN IN FIG. 5.5.12c.

5.12) FROM THE RESULTS IN EXAMPLE 5.2.2 A STABLE OSCILLATION OCCURS AT

$$\omega_0 = \frac{1}{\sqrt{L_0 C_0}}$$

AT THE AMPLITUDE $A' = A'_0$, DETERMINED BY

$$G_{IN}(A'_0) + G_0 = 0$$

THE OSCILLATION IS STABLE (SEE EXAMPLE 5.2.2) BECAUSE

$$\left.\frac{\partial G_{IN}(A')}{\partial A'}\right|_{A'=A'_0} \left.\frac{dB_0(\omega)}{d\omega}\right|_{\omega=\omega_0} > 0$$

WHERE $B_0(\omega) = j\omega C_0 + \frac{1}{j\omega L_0}$

THE START OF OSCILLATION CONDITION REQUIRES THAT

$$|G_{IN}(0)| + G_0 > 0$$

5.13) $Z_L = 8.5 - j24\,\Omega$, $z_L = \frac{Z_L}{50} = 0.17 - j0.48$, $y_L = 0.656 + j1.85$

THE MATCHING USED IS SHOWN IN THE Y SMITH CHART. THE MOTION FROM THE ORIGIN TO "a" IS IMPLEMENTED WITH A SHORT-CIRCUITED SHUNT STUB. THE MOTION FROM "a" TO "b" IS IMPLEMENTED WITH A SERIES TRANSMISSION LINE OF LENGTH 0.018λ. THE MOTION FROM "b" TO y_L (i.e., ALONG A CONSTANT CONDUCTANCE CIRCLE) IS IMPLEMENTED WITH A CAPACITOR IN SHUNT. THIS CAPACITOR CAN BE OBTAINED USING A VARACTOR DIODE. THE VALUE OF THE CAPACITOR IS:

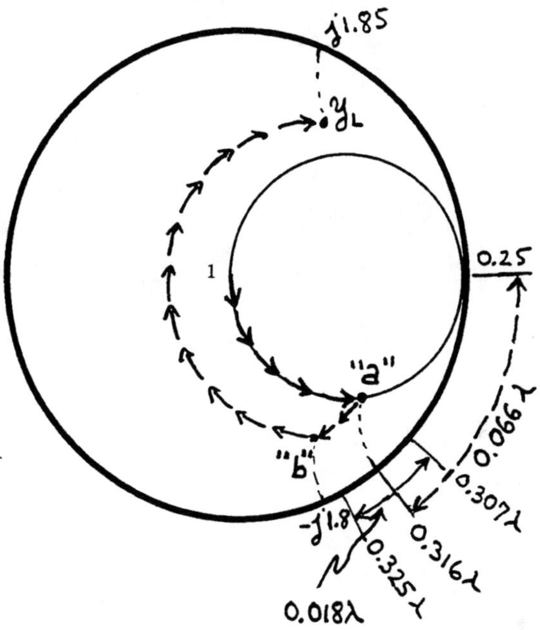

$$y_c = j1.85 - (-j1.8) = j3.65$$

$$\therefore C = \frac{3.65/50}{2\pi(2.75\,10^9)} = 4\text{ pF}$$

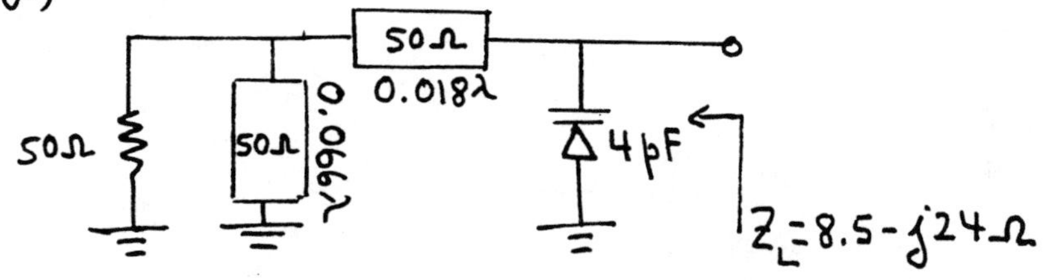

5.14) IT IS A POTENTIALLY UNSTABLE TRANSISTOR

OUTPUT STABILITY CIRCLE: $C_L = 0.921\angle{-76.1^\circ}$, $r_L = 0.858$

SEVERAL VALUES OF Γ_T IN THE UNSTABLE REGION SHOULD BE TRIED, AND THE ASSOCIATED Γ_{IN} CALCULATED. FOR THIS DESIGN WE USED $\Gamma_T = 0.7\angle{-130^\circ}$, THEN $\Gamma_{IN} = 2.09\angle{-136^\circ}$ AND $Z_{IN} = -20 - j17.2\,\Omega$.

LET $Z_L = \frac{20}{3} + j17.2 = 6.7 + j17.2$ OR $\Gamma_L = 0.786\angle{141.5^\circ}$

AN IMPLEMENTATION OF THE OSCILLATOR IS:

THE CAPACITOR ($C = 2.3$pF) IS IMPLEMENTED WITH A VARACTOR DIODE).

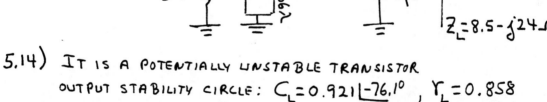

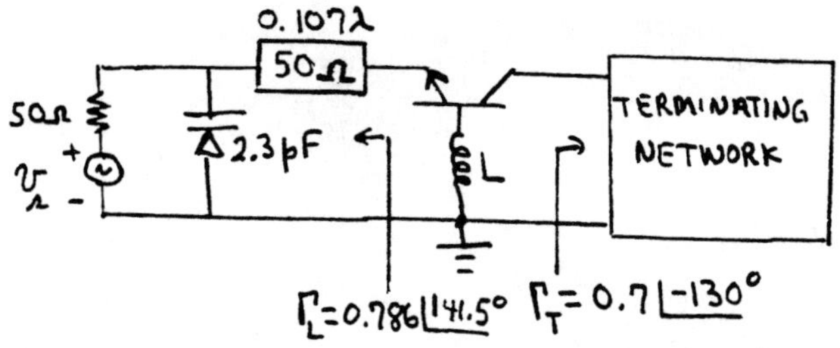

5.15) (a) $P_{OUT} = P_{sat}\left(1 - e^{-G_o P_{IN}/P_{sat}}\right)$ (1)

MAXIMUM OSCILLATOR POWER OCCURS WHEN $P_{OUT} - P_{IN}$ IS A MAXIMUM, OR $\dfrac{\partial P_{OUT}}{\partial P_{IN}} = 1$. SINCE

$$\frac{\partial P_{OUT}}{\partial P_{IN}} = P_{sat}\left(-\frac{G_o}{P_{sat}}\right)\left(e^{-G_o P_{IN}/P_{sat}}\right) = G_o e^{-G_o P_{IN}/P_{sat}} = 1$$

THEN,

$$e^{-G_o P_{IN}/P_{sat}} = \frac{1}{G_o} \quad \text{OR} \quad P_{IN} = P_{sat}\frac{\ln G_o}{G_o} \qquad (2)$$

SUBSTITUTING (2) INTO (1), P_{OUT} CAN BE EXPRESSED AS:

$$P_{OUT} = P_{sat}\left(1 - \frac{1}{G_o}\right) \qquad (3)$$

FROM (3) AND (2), THE MAXIMUM OSCILLATOR POWER ($P_{osc}(max)$) IS

$$P_{osc}(max) = P_{OUT} - P_{IN} = P_{sat}\left(1 - \frac{1}{G_o} - \frac{\ln G_o}{G_o}\right)$$

(b) $G_o = 7.5\,dB$ OR 5.623

$P_{sat} = 1\,W$

$$P_{osc}(max) = 1\left(1 - \frac{1}{5.623} - \frac{\ln 5.623}{5.623}\right) = 0.515\,W$$

(c)

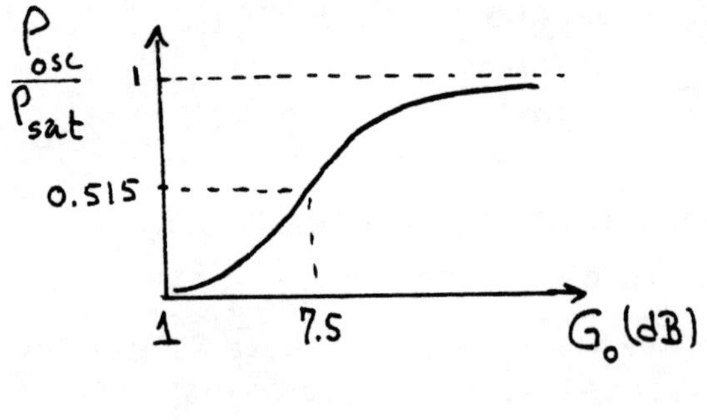